System-on-Chip Security

Farimah Farahmandi • Yuanwen Huang
Prabhat Mishra

System-on-Chip Security

Validation and Verification

Springer

Farimah Farahmandi
University of Florida
Gainesville, FL, USA

Yuanwen Huang
Google, Mountain View
CA, USA

Prabhat Mishra
University of Florida
Gainesville, FL, USA

ISBN 978-3-030-30598-7   ISBN 978-3-030-30596-3   (eBook)
https://doi.org/10.1007/978-3-030-30596-3

This Springer imprint is published by the registered company Springer Nature Switzerland AG.
The registered company address is: Gewerbestrasse 11, 6330 Cham, Switzerland

To our families

Preface

We are living in a connected world, where a wide variety of computing and sensing components interact with each other. Secure computation and trusted communication are essential as intelligent computing devices are increasingly embedded in every possible device in our daily life such as wearable devices, autonomous vehicles, and smart homes. Any failure of security and trust requirements of these devices may endanger human life and environment by causing damages to critical infrastructure, violating personal privacy, or undermining the credibility of a business. Attacks on hardware can be more critical than traditional attacks on software since patching is extremely difficult (almost impossible) on hardware designs. Note that hardware designs are fixed after fabrication, and any existing vulnerability in their implementations can be exploited by attackers. Moreover, the same attack can be repeated on every instance of the fabricated design. The problem gets worse when we consider the increasing complexity of the semiconductor designs. Therefore, effective and well-developed hardware security validation and verification techniques are vital for the security assurance of today's designs.

Modern computing devices are designed using System-on-Chip (SoC) technology. Modern SoC designs contain several highly sensitive assets such as encryption keys, device configurations, and on-device protected data that are responsible for keeping our personal, financial, and intimate physiological information safe and secure. These assets should be protected from any unauthorized access. With the globalization of the semiconductor industry, the outsourcing and integration of third-party hardware Intellectual Property (IP) has become a common practice in SoC design methodology. However, it raises significant security concerns as an attacker can insert malicious components (e.g., hardware Trojans) in third-party IPs and tamper the system. There are a wide variety of security vulnerabilities for hardware designs. Attacks on hardware can be immensely dangerous and can harm human life and environment by causing damages to critical infrastructure, violating personal privacy, or undermining the credibility of a business. These attacks may arise from a wide variety of sources such as malicious components, insecure connection to software, firmware, and other devices as well as side-channel vulnerabilities through energy, power, and performance profiles. Given the importance of hardware security and the extreme consequences of vulnerable SoC designs, it is critical to ensure their correctness from both functional and security perspectives. There has been a plethora of research (including conference and journal publications) in the last decade on developing efficient security validation and verification techniques. This book covers a wide variety of state-of-the-art SoC security validation and verification techniques.

This book provides a comprehensive overview of SoC security vulnerabilities and corresponding security verification techniques based on formal verification, machine learning, simulation-based validation, and side-channel analysis. These techniques are applicable across different design abstraction levels to address both detection and localization (mitigation) of SoC security vulnerabilities. The presentation of topics has been divided into five categories with each category focusing on a specific aspect of the big picture. A brief outline of the book is provided below.

1. **SoC Security Validation Preliminaries:** The first part of the book includes three introductory chapters on SoC security validation and related topics.

   - Chapter 1 introduces the modern semiconductor supply chain and provides an overview of SoC security vulnerabilities.
   - Chapter 2 describes the fundamental challenges in validation and verification of SoC security and trust. Specifically, it highlights why existing functional validation methodology is enough for SoC security verification.
   - Chapter 3 presents SoC trust metrics and benchmarks that are vital in evaluating the quality of security validation techniques as well as the trustworthiness of SoC designs.

2. **Security Verification Using Formal Methods:** The second part of the book focuses on formal and semi-formal approaches for SoC security verification.

   - Chapter 4 presents an equivalence checking framework using symbolic algebra to identify anomalies in the implementation compared to the golden reference model.
   - Chapter 5 describes an efficient framework for detection and localization of hardware Trojans using symbolic algebra.
   - Chapter 6 outlines an automated approach for detecting security vulnerabilities in finite state machines.
   - Chapter 7 presents SoC trust validation techniques using security properties.

3. **Security Validation Using Simulation and Learning Techniques:** The third part of the book deals with security validation techniques using simulation-based validation as well as machine learning.

   - Chapter 8 describes automated test generation techniques for detection of malicious implants (e.g., hardware Trojans).
   - Chapter 9 provides an overview of hardware Trojan detection techniques using machine learning.

4. **Security Validation Using Side-Channel Analysis:** The fourth part of the book looks at SoC security validation techniques using side-channel signatures such as dynamic current and path delay.

   - Chapter 10 describes hardware Trojan detection techniques using dynamic current-based side-channel analysis.
   - Chapter 11 presents efficient techniques for detection of hardware Trojans using path delay analysis.

5. **Conclusions and Future Directions:** The last part concludes the book with a summary and discussion on future directions.

   - Chapter 12 concludes the book with an executive summary as well as discussion on security validation challenges of future SoCs.

We hope you enjoy reading this book and find the information useful for applying SoC security validation and verification techniques in designing secure and trustworthy systems.

Gainesville, FL, USA   Farimah Farahmandi
June 30, 2019   Yuanwen Huang
                                                                                          Prabhat Mishra

Acknowledgements

This book is the result of a decade long academic research and industrial collaborations. The book includes the hardware security validation techniques and insights that resulted from Ph.D. dissertations of Dr. Farimah Farahmandi and Dr. Yuanwen Huang. We would like to acknowledge our sponsors for providing the financial support to enable this research work. This work was partially supported by National Science Foundation (CNS-1441667), Semiconductor Research Corporation (2014-TS-2554), and Cisco Systems. We would like to acknowledge the contributions of Prof. Fareena Saqib (UNC Charlotte) and Prof. Jim Plusquellic (University of New Mexico) for contributing the book chapter on hardware Trojan detection schemes using path delay and side-channel analysis (Chap. 11).

Contents

Acronyms

ABV	Assertion based validation
ATPG	Automatic test pattern generator
BDD	Binary decision diagrams
BMC	Bounded model checking
CNF	Conjunctive normal form
DUV	Design under validation
FSM	Finite state machine
IC	Integrated circuit
IP	Intellectual property
LTL	Linear temporal logic
PSL	Property specification language
RTL	Register transfer level
SAT	Satisfiability
SoC	System-on-Chip
SVA	System-Verilog Assertion
TLM	Transaction level modeling

Chapter 1
System-on-Chip Security Vulnerabilities

1.1 Introduction

There is a new trend toward validation of complex computing systems, which is hardware security verification and validation. Previously, hardware systems were considered secure, trusted, and static where every other computing components (such as firmware, hypervisors, operating systems, user applications) were built over them. However, hardware cannot be considered as root-of-trust anymore as recent research practices [16, 17] have shown that hardware systems can be as vulnerable as software systems toward security attacks. The importance of hardware security validation significantly increases when considering Internet-of-Things (IoT) devices. Highly complex, connected, and smart IoT devices are increasingly embedded in our daily life (almost everywhere) and they are recording, analyzing, and communicating some of our most intimate personal information in order to improve the quality of our lives. The core computing functionality of each of these IoT devices is performed by one or more complex System-on-Chip (SoC) designs. It is a significant challenge to verify the security requirements of SoCs in IoT devices, primarily due to increasing design complexity coupled with shrinking time-to-market constraints. Verification is already a major bottleneck in modern chip design life cycle where more than 70% of the resources and engineering time are spent on verification efforts [6] to ensure the correct functionality, performance, timing, and reliability of a hardware design. The verification problem becomes more challenging to ensure SoCs are secure and trusted and operate in compliance with their specifications, especially when considering the security requirements of diverse applications and evolving use cases of IoT devices. In the absence of comprehensive SoC security verification, vulnerable IoT devices can lead to catastrophic consequences ranging from violating personal privacy, hurting the reputation of a business to endangering human lives. Therefore, detecting and locating hardware Trojans are extremely challenging due to their stealthy behavior and it requires the development of efficient and scalable security validation approaches. Developing

© Springer Nature Switzerland AG 2020
F. Farahmandi et al., *System-on-Chip Security*,
https://doi.org/10.1007/978-3-030-30596-3_1

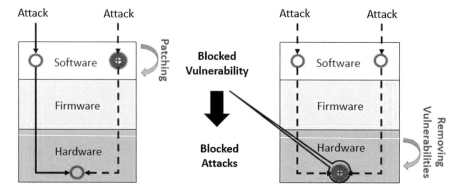

Fig. 1.1 Hardware vulnerabilities enable different attacks (software and firmware-based attacks). Blocking hardware vulnerabilities will address current and future security issues [5]

efficient and well-designed hardware security validation approaches is an essence to create more secure and trustworthy IoT devices, and hardware systems should be verified and validated against various security and trust requirements before integration in computing systems.

Existing hardware security verification approaches are often ad hoc and manual (i.e., rely on human ingenuity and experience). There is a critical need to identify all possible security vulnerabilities and fix them using automatic and reliable mechanisms during security validation. Attacks on hardware can be more critical and destructive than traditional software attacks since patching is extremely difficult (almost impossible) on hardware designs. Moreover, a security attack can be successfully repeated on every instance of a vulnerable IoT device. As shown in Fig. 1.1, hardware-level vulnerabilities are extremely important to be fixed before deployment since it affects the overall system security. Based on common vulnerability exposure (CVE-MITRE) estimates, if hardware-level vulnerabilities are removed, the overall system vulnerability will reduce by 43% [4, 5].

An SoC is an integrated circuit that encompasses all components of a computing system such as processing units, memory, secondary storage, input/output ports in a single chip [15]. An SoC typically contains several security assets and sensitive information such as encryption keys, Original Equipment Manufacturer (OCM) and Original Component Manufacturer (OEM) keys, developer keys, digital rights management (DRM) keys, and configuration bits that are needed to be protected from adversaries [12]. An SoC is usually constructed from several pre-designed intellectual property (IP) blocks. Each IP is responsible to implement a specific functionality (e.g., CPU, memory units, memory controller, analog-to-digital converter, digital-to-analog converter, digital signal processing unit, etc.) as well as communicate with other IPs through standard communication fabrics such as network-on-chip (NoC). As shown in Fig. 1.2, a typical SoC may also come with various security IPs such as crypto (encryption and decryption) cores, True Random Number Generator (TRNG) modules, Physical Unclonable Function (PUF) units,

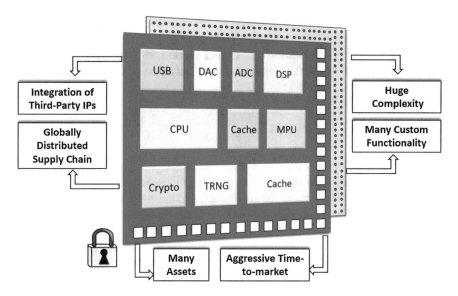

Fig. 1.2 An SoC design integrates a wide variety of IPs in a chip. It can include one or more processor cores, on-chip memory, digital signal processor (DSP), analog-to-digital (ADC) and digital-to-analog converters (DAC), controllers, input/output peripherals, and communication fabric. Huge complexity, many custom designs, distributed supply chain, and integration of untrusted third-party IPs make security validation challenging

one-time memory blocks, etc. The security IPs either generate, propagate, use, or manage assets during runtime. Therefore, security assets are distributed at different IPs across the SoC and they should be protected using security policies. However, there are many security vulnerabilities that can be exploited by attackers, which could compromise the security of SoCs by leaking sensitive information, tampering the functionality of the design, or causing a denial of service.

The IP-based SoC design methodology is a popular approach since it provides an opportunity for IP reusing, which leads to reducing design costs, as well as meeting time-to-market constraints. With the globalization of the IC industry, IP outsourcing and integration has become a trend for SoC design [1]. However, it raises significant security concerns as the attacker can insert malicious modifications in third-party IPs and tamper the system. Additionally, assets can be leaked through side-channel information and existing vulnerabilities in IPs. Security vulnerabilities can be inserted intentionally or introduced unintentionally at different stages of SoC design, such as in the high-level specifications (e.g., transaction-level modeling, TLM, and register transfer level, RTL, models), synthesized gate-level netlist, layout, as well as in the fabricated chip by an attacker. In this book, we show various threat models for SoC designs as well as their IPs and we discuss several verification and validation approaches to detect various security vulnerabilities in them. We also show different mitigation techniques to address them.

The rest of this chapter is organized as follows: Sect. 1.2 presents the source of hardware security attacks at different stages of a design life cycle. Section 1.3 describes the security vulnerabilities (threat model) in the current SoC design methodology. Finally, Sect. 1.4 describes the organization of this book.

1.2   Sources of Attacks in SoCs

Security threats can be introduced throughout the IC design, as well as the manufacturing process. In the pre-silicon stage, vulnerabilities can be introduced due to (1) designer mistakes, rogue employees, and untrusted third-party IPs during the design integration phase; (2) untrusted electronic design automation (EDA) tools in the synthesis phase; (3) untrusted EDA tools and untrusted vendors when design-for-test (DFT), design-for-debug (DFD), and dynamic power management (DPM) functions are added. In the post-silicon stage, vulnerabilities can come from (1) untrusted foundry during manufacturing, and (2) physical attacks or side-channel attacks after the chip is shipped. An SoC design can encounter security threats during different stages of its life cycle, as shown in Fig. 1.3. We have listed the sources of attacks in SoCs as follows.

1.2.1   Design Stage

Design of an SoC starts with defining the high-level behavior and requirements using natural languages, as well as high-level languages such as C and C++. Hardware designers implement specifications using RTL descriptions. In the past, all of the components of an SoC are designed in-house. However, due to constraints on time-to-market and exponential increase of design complexity, outsourcing and integration of third-party hardware IPs have become a common practice for SoCs. Attacks in the design stage can occur through the integration of third-party IPs. These IPs may come with deliberate malfunctions that pose significant security threats to the security of SoCs. Malfunctions may leak secret information to adversaries or reduce the reliability of the design. These malfunctions can also be introduced using rogue designers (insider attacks). Insider threats are particularly dangerous since they have full observability and access to the whole design and source files. Moreover, IP theft can also happen at the design stage. Stolen IPs will lead to loss of venue for the IP owner and producing counterfeit instances of the design. Furthermore, analyzing of stolen IPs will help to find existing vulnerabilities of the design, as well as new ways (from software or hardware) to attack the SoC.

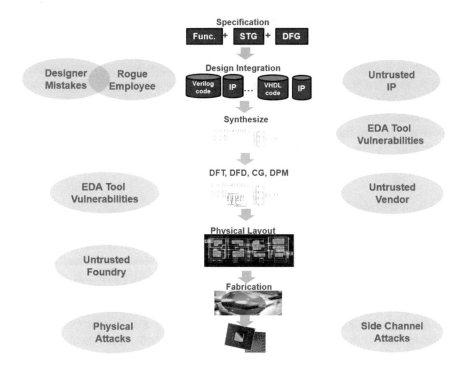

Fig. 1.3 Potential threats during SoC design flow. It shows various types of threats (represented by ovals) during different design stages: specification, integration, synthesis/DFT/DPM, layout, and fabrication

1.2.2 Synthesis RTL to Layout

When the SoC design and integration is done at RTL, the design is needed to be synthesized to a gate-level netlist. The synthesis process is done using third-party EDA tools (e.g., Synopses Design Compiler [14], Cadence Genus Synthesis Solution [2], etc.). These tools only take timing, performance, area, and power into consideration, and they are often unaware of security while transforming the design to the lower levels of abstraction. While performing design optimization, these tools may create unintentional vulnerabilities into the design. One example of such vulnerabilities is that while the synthesis tool tries to optimize the controller design, it may introduce don't care states in finite state machines (FSMs). The assumption was that these additional states are not accessible through states transitions (from the initial state of the FSM or other stats) and they do not affect the correct functionality of the design. However, recent studies have shown that these states are reachable through fault injection attacks [10, 11]. Now if the don't care states are connected to protected states (those states that control security-critical operations) of the design, an adversary can inject faults to access those don't care stats and access to the protected states illegally.

The gate-level netlist is required to be mapped to standard cell library and transistor-level netlist (layout). The gate-level netlist and the layout of design can go to untrusted venues for different purposes such as DFD and DFT insertion, power optimization, clock-tree insertion, etc. Since those entities have write access to the netlist, they can inject malicious functionality in the design by adding/removing gates and transistors or manipulating the interconnects of the layout. Moreover, these entities can reverse engineer the netlist and create IP piracy and counterfeit problems. Having full knowledge of the design will also lead to extra information that facilitates new attacks.

1.2.3 Fabrication and Manufacturing

When the layout of the design is finalized, it will be sent to the foundry to fabricate the chips. Due to the increased cost of fabrication, design houses send their designs to potentially untrusted foundries. Attackers in the foundry can add malicious functionality into the chip. IP piracy and reverse engineering of the design to create counterfeits also can happen with an untrusted foundry. An untrusted foundry can introduce overproduction threat. The foundry may not honor the number of chips stated in the contract and creates more chips and makes profits out of them by selling them in the black market.

1.2.4 In-Field Attacks

When a chip is deployed into the final design, it will be susceptible to various types of attacks. If a Trojan was inserted during design, synthesis, or fabrication stages, it can be activated to perform the intended attack or malfunction. The malfunction can also be activated by injecting faults in the design (using changing the clock frequency, voltage, local heating, intensive light pulses, etc.). An attacker can monitor physical characteristics of a design (such as delays, power consumption, transient current leakage) to recover secret information. Moreover, a well-equipped attacker can perform reverse engineering through depackaging, delayering, and imaging of the chip to extract information about the design and enable IP theft and counterfeiting. Moreover, high-precision and nondestructive probing equipment can be used to obtain secret information (e.g., different keys that are stored in non-volatile memories). Last but not least, refurbished and recycled chips may be presented as new chips. It is a dangerous threat especially when the functionality of a system is dependent on those chips since the system may not be reliable or come with permanent faults/failures (refurbished chips did not pass some manufacturing tests).

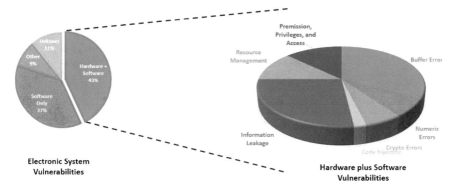

Fig. 1.4 Different categories of common hardware and software vulnerability exposure in an electronic system [5]

It is of paramount importance to verify the trustworthiness of an SoC. In order to trust a design, security verification and debugging should be done at each of the stages.

1.3   Threat Model

In this section, we talk about different threat models that endanger the security of SoCs. As shown in Fig. 1.4, the potential SoC vulnerabilities would be huge once we consider seven classes of hardware security vulnerabilities (access privileges, buffer errors, resource management, information leakage, numeric errors, crypto errors, and code injection) coupled with software and firmware attacks that threaten the security and integrity of the design [5]. Therefore, detecting and locating these vulnerabilities are extremely challenging due to their stealthy behavior, and it requires efficient and scalable security validation approaches to be developed. Each design should be verified against all of these threat models to ensure the correct and secure behavior of the design. In terms of hardware security verification, we categorized them into four classes: hardware Trojans, access violations, fault injection attacks, and side-channel leakage. In this section, we briefly describe each of these threat models.

1.3.1   Hardware Trojans

Hardware Trojans are malicious modifications of an integrated circuit which are designed to disable or bypass the security of design. They can also create denial of service by tampering the functionality of the design. Hardware Trojans are

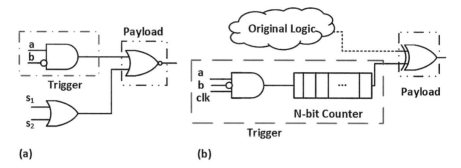

Fig. 1.5 Simple examples of hardware Trojans. (**a**) A combinational Trojan that can be triggered using rare condition $a = 1$ and $b = 0$. (**b**) A sequential Trojan that is triggered when rare condition $a = 1$ and $b = 0$ happens 2^N times, where N is the length of the counter. The effect of Trojans can be propagated using payload circuit

hard-to-detect malicious components which are inactive for most of the execution time and they can be activated under rare input conditions which trigger them. Hardware Trojans can be inserted in an SoC by integration of untrustworthy IPs gathered from third parties, internally by a rouge employee or by EDA tools. Trojans can be designed as a combinational circuit (e.g., a k-bit comparator) or a sequential circuit (e.g., k-bit counter) as shown in Fig. 1.5. Usually, hardware Trojans consist of two parts: trigger and payload. The trigger is responsible for checking the activation conditions and the payload is the entire activity of the Trojan and it is responsible for propagating the effect of the Trojan when it becomes activated. Trojans can be activated by change of functionality (digital conditions) or change in the physical characteristics of the design (analog conditions) such as temperatures. Smart adversaries design hardware Trojans such that they are unlikely to arise during normal testing and validation stages (to create a stealthy behavior) but Trojans can be activated after long hours of in-field execution.

A major challenge for Trojan identification is that Trojans are usually stealthy [1]. It is difficult to construct a fault model to characterize Trojan's behavior. Moreover, Trojans are designed in a way that they can be activated under very rare conditions and they are hard-to-detect. Therefore, it is difficult to activate a Trojan and even more difficult to detect or locate it. As a result, conventional validation methods are impractical to detect hardware Trojans. Conventional structural and functional testing methods are not effective to activate trigger conditions since there are many possible Trojans and it is not feasible to construct a fault model for each of them. As a result, existing EDA tools are incapable of detecting hardware Trojans to differentiate between trustworthy third-party IPs and untrusted ones. Furthermore, malicious hardware can easily bypass traditional software-implemented defense techniques as it is a layer below the entire software stack. Similarly, a benign debug/test interface or uncharacterized parametric behavior can be exploited by an adversary during legitimate in-field debug.

Trojans can be inserted into hardware design using various ways as listed below:

- **Rare Nodes:** A smart adversary tries to design trigger conditions such that they are satisfied in very rare situations and usually after long hours of operation [3]. Rare conditions at internal nodes (rare nodes) are candidates for hiding the malicious functionality. Figure 1.5 shows examples of hardware Trojans that are designed using rare nodes. Figure 1.5a shows a combinational Trojan whose trigger is dependent on a set of rare nodes ($a = 1$ and $b = 0$). The Trojan will be activated when the respective conditions on rare nodes are satisfied. On the other hand, Fig. 1.5 shows a simple sequential Trojan which is triggered by the overflow output of a counter. The counter increments when it is enabled using conditions ($a = 1$ and $b = 0$) and the Trojan is activated when a series of $a = 1$ and $b = 0$ events happens until the counter reaches a specific value.
- **Rare Branches:** An adversary (e.g., a rogue designer or an untrusted IP vendor) can insert hard-to-detect Trojans in the RTL design and hide them under rare branches and continuous/concurrent assignments. Otherwise, traditional simulation techniques using random or constrained-random tests can detect them, and the attacker's attempt would fail.
- **Gate Misplacement:** Any deviation from the specification may endanger the correct functionality, trustworthiness, and the security of the design. Notably, gate replacement errors in the gate-level netlist can change the correct functionality of design and insert anomaly in its implementation. Moreover, gate replacement error may pose security threats since it can act as a bit-flip (in comparison with the golden behavior) and cause unauthorized transitions to protected states of the design, wrong results, and denial of service. Gate replacement anomalies are small malicious modifications and have negligible effect on physical characteristics (area, power, and energy) of the design. Therefore, they cannot be detected during design review. Moreover, they cannot be easily activated using random and constraint-random validation approaches.

A design should be verified and validated comprehensively to ensure that there is no Trojan or malfunction inserted in it. In this book we cover several methods such as formal methods (Chaps. 4, 5, and 7), simulation and learning-based approaches (Chaps. 8 and 9), and side-channel analysis (Chaps. 10 and 11).

1.3.2   Access Violation

Critical data as well as protected states of the design should only be accessed by authorized sources as any unauthorized access mechanisms can lead to illegal read/write of assets or changing the flow of the program to bypass the security of the overall system and threaten its integrity. Therefore, the content of memory locations (e.g., instruction and data cache units, registers, RAM blocks, and hard drives) with sensitive information and assets should be protected from unauthorized modifications. Violation of memory confidentiality may allow adversaries to achieve their goal without providing correct input and lead to a leak of sensitive information. For example, the system's assets/sensitive computations may be accessed through DFT and DFD infrastructure for legitimate debugging reasons and in order to facilitate hardware post-silicon validation. However, an attacker should not be able to access those information while the chip is in functional mode. It is also important to check how the memory is accessed in order to prevent vulnerabilities like buffer overflow and integer overflow attacks. Buffer overflows can lead to overwrites in adjacent memory locations and cause integrity problems. Recently it has been shown that integer/buffer overflow attacks as well as unauthorized accesses can happen due to speculative components of the hardware design such as exception handler unit and branch predictor that allow programs to steal the secret stored in the memories that they are not allowed to access [8, 9]. Security validation approaches should check all access paths to critical information and memory location of the design. Designers need to block those detected unprivileged accesses. Chapters 6 and 7 present security validation mechanisms for such vulnerabilities.

1.3.3 Fault Injection Attacks

Over the past decade, fault injection attacks have grown from a crypto-engineering curiosity to a systematic adversarial technique [19]. FSMs in the control path are also susceptible to fault injection attacks, and the security of the overall SoC can be compromised if the FSMs controlling the SoC are successfully attacked. For example, it has been shown that the secret key of the RSA encryption algorithm can be detected when FSM implementation of the Montgomery ladder algorithm is attacked using fault injection [13]. Fault injection attacks can be performed by changing the clock frequency (overclocking), reducing the voltage, and heating the device to violate the setup time constraint of state flip-flops to bypass a normal state transition and enter a protected state. The non-uniform path delay distribution of an FSM enables an attacker to violate setup time of certain flip-flops and bypass the security of the design. For fault injection attacks, the adversary should have physical access to the device. Setup time violations can be performed by different fault injection methods, including overclocking, reducing the voltage, and/or heating the device [18]. To prevent fault injection attacks on FSMs, it is critical to identify and remove FSM vulnerabilities. The susceptibility to fault injection attacks should be analyzed in both datapath as well as control logic of an SoC. For datapath, we should check the likelihood of creating timing violation faults and if the fault will propagate throughout the design. For control logic, the state transition graph of the controller circuit should be checked to see if an adversary can cause timing violation to bypass normal state transitions and get access to state which causes security vulnerability. In Chap. 6, we propose a technique to formally detect such vulnerabilities. The challenges in trust validation of controller designs come from the fact that we need to detect illegal accesses to the design states in addition to verifying legal transitions. The state-space of this problem is exponential.

1.3.4 Side-Channel Attacks

Timing information, power consumption, electromagnetic emanation, and even sound of a design can be extracted by an attacker to gain more information about the design and be able to attack. For example, an attacker can guess some internal values or secret keys by measuring the execution time of various computations (note that "0" or "1" bits in a register can initiate different operations). Extracting side-channel information may require some knowledge about the internal structure of the design. However, some of these attacks such as differential power side-channel attacks [7] are black-box attacks. Unfortunately, side-channel analysis has a common issue, i.e., the sensitivity of side-channel signatures is susceptible to thermal and process variations. Therefore, the success of these attacks is determined by the quality and precision of equipment that are used for measurement.

Power-side channel attacks use the amount of power consumption and transient/dynamic current leakage to attack the design. A device like an oscilloscope can be used to collect power traces, and those traces are statistically analyzed using correlation analysis to derive secret information of the design. Therefore, it is very important to develop automated security validation methods that can identify power side-channel leakage. We need to detect the parts of a design that is responsible for power side-channel leakage in an automated fashion. Chapter 10 presents techniques to detect these vulnerabilities.

Any implicit or explicit control flow that depends on the asset value can create side-channel timing leakage and make the design vulnerable to timing attacks. To remove timing side-channel attacks, the security verification tools need to ensure that the execution time is independent of the asset value. The assets dependent control flows make the design vulnerable to timing side-channel leakage. Chapter 11 covers security validation methods to detect such vulnerabilities.

1.4 Book Organization

In this book, we provide different security verification and validation approaches to identify security and trust vulnerabilities that are introduced at different stages of the design. These techniques are based on formal methods, simulation-based approaches, machine learning, and side-channel analysis. These techniques can be applied at IP-level, pre-silicon design (after integration of soft IPs), and post-silicon. The organization of this book is as follows:

- Chapter 2 presents the fundamental challenges in verifying SoC security vulnerabilities. Specifically, it also highlights the limitations of applying the existing functional and security validation methods.
- Chapter 3 describes security metrics and benchmarks (both dynamic and static) which are necessary for evaluating the trustworthiness of SoCs as well as measuring the effectiveness of any security verification/validation technique.
- Chapter 4 describes an automated methodology for anomaly detection in complex arithmetic circuits. It used the remainder produced by equivalence checking methods to generate directed tests as well as fixing the security vulnerabilities. The threat model is considered malfunction insertion using gate misplacement.
- Chapter 5 presents an automated approach to localize hardware Trojans in third-party IPs using symbolic algebra. This chapter considers hardware Trojans that change the functionality of the design (e.g., add additional malfunction) as the threat model.
- Chapter 6 highlights the importance of securing FSMs against fault injection attacks and access violations. This chapter presents a formal approach to detect anomalies using symbolic algebra. This chapter also discusses some design rules to avoid such vulnerabilities.
- Chapter 7 discusses the importance of developing security properties that allow detection of security violations such as information leakage at the early stages of the design cycle.
- Chapter 8 focuses on efficient simulation-based validation approaches as well as test generation techniques for hardware Trojan detection.
- Chapter 9 presents machine learning techniques as well as feature extraction techniques for the detection of hardware Trojans.
- Chapter 10 discusses a side-channel analysis framework based on current and power signatures to detect malfunctions in an SoC.
- Chapter 11 surveys a wide variety of delay-based side-channel analysis approaches for detection of side-channel security vulnerabilities. It describes a wide range of timing and power analysis techniques to detect hardware Trojans.
- Chapter 12 provides a summary of techniques covered in the book. This chapter also highlights the future directions for security verification and validation of an SoC.

1.5   Summary

This chapter introduced the modern semiconductor supply chain and provided an overview of SoC security vulnerabilities. Specifically, it highlighted various types of potential threats during different design stages. It provided an overview of multiple SoC security vulnerabilities. This chapter presented the fact that SoC designs are required to be validated/verified from security and trust aspects.

References

1. S. Bhunia, M.S. Hsiao, M. Banga, S. Narasimhan, Hardware Trojan attacks: threat analysis and countermeasures. Proc. IEEE **102**(8), 1229–1247 (2014)
2. Cadence Genus Synthesis Solution, https://www.cadence.com/content/cadence-www/global/en_US/home/tools/digital-design-and-signoff/synthesis/genus-synthesis-solution.html
3. R.S. Chakraborty, F. Wolf, C. Papachristou, S. Bhunia, MERO: a statistical approach for hardware Trojan detection, in *International Workshop on Cryptographic Hardware and Embedded Systems (CHES'09)* (2009), pp. 369–410
4. Common Weakness Enumeration, https://cwe.mitre.org
5. DARPA System Security Integrated Through Hardware and Firmware (SSITH), https://www.fbo.gov/index?s=opportunity&mode=form&id=ea2550cb0c42eb91c7292377824a58b7
6. H. Kaeslin, *Top-down Digital VLSI Design: From Architectures to Gate-level Circuits and FPGAs* (Morgan Kaufmann, Waltham, 2014)
7. P.C. Kocher, J. Jaffe, B. Jun, Differential power analysis, in *Proceedings of the 19th Annual International Cryptology Conference on Advances in Cryptology, Series CRYPTO '99, London, UK* (Springer, London, 1999), pp. 388–397. [Online]. Available: http://dl.acm.org/citation.cfm?id=646764.703989
8. C. Li, J. Gaudiot, Online detection of spectre attacks using microarchitectural traces from performance counters, in *2018 30th International Symposium on Computer Architecture and High Performance Computing (SBAC-PAD), Lyon, France* (2018), pp. 25–28
9. M. Lipp, M. Schwarz, D. Gruss, T. Prescher, W. Haas, A. Fogh, J. Horn, S. Mangard, P. Kocher, D. Genkin, Y. Yarom, M. Hamburg, Meltdown: reading kernel memory from user space, in *27th Security Symposium (USENIX Security)* (2018), pp. 973–990
10. A. Nahiyan, K. Xiao, K. Yang, Y. Jin, D. Forte, M. Tehranipoor, AVFSM a framework for identifying and mitigating vulnerabilities in FSMs, in *2016 53nd ACM/EDAC/IEEE Design Automation Conference (DAC)* (IEEE, Piscataway, 2016), pp. 1–6
11. A. Nahiyan, F. Farahmandi, P. Mishra, D. Forte, M. Tehranipoor, Security-aware FSM design flow for identifying and mitigating vulnerabilities to fault attacks. IEEE Trans. Comput. Aided Des. Integr. Circuits Syst. **38**(6), 1003–1016 (2019)
12. S. Ray, E. Peeters, M.M. Tehranipoor, S. Bhunia, System-on-chip platform security assurance: architecture and validation. Proc. IEEE **106**(1), 21–37 (2018)
13. B. Sunar, G. Gaubatz, E. Savas, Sequential circuit design for embedded cryptographic applications resilient to adversarial faults. IEEE Trans. Comput. **57**(1), 126–138 (2008)
14. Synopsis Design Compiler, https://www.synopsys.com/implementation-and-signoff/rtl-synthesis-test.html
15. System on a chip, https://en.wikipedia.org/wiki/System_on_a_chip
16. M. Tehranipoor, F. Koushanfar, A survey of hardware Trojan taxonomy and detection. IEEE Des. Test Comput. **27**(1), 10–25 (2010)
17. M. Tehranipoor, C. Wang, *Introduction to Hardware Security and Trust* (Springer Science & Business Media, New York, 2011)
18. B. Yuce, N.F. Ghalaty, P. Schaumont, TVVF: estimating the vulnerability of hardware cryptosystems against timing violation attacks, in *2015 IEEE International Symposium on Hardware Oriented Security and Trust (HOST)* (IEEE, Piscataway, 2015), pp. 72–77
19. B. Yuce, N.F. Ghalaty, C. Deshpande, C. Patrick, L. Nazhandali, P. Schaumont, Fame: fault-attack aware microprocessor extensions for hardware fault detection and software fault response, in *Proceedings of the Hardware and Architectural Support for Security and Privacy 2016* (ACM, New York, 2016), p. 8

Chapter 2
SoC Security Verification Challenges

2.1 Introduction

Verification/validation is a significant bottleneck in the design steps of a System-on-Chip (SoC) such that it consumes more than 70% of design efforts. Significant validation efforts come from the fact that an SoC should be verified against several objectives, such as correct functionality, timing, power, energy consumption, reliability, and security in pre- and post-silicon stages, before it can be used in hardware devices. Moreover, an SoC has various working domains, including digital, analog, and mixed-signals. All components in different areas should work with each other correctly to create the expected behavior. Therefore, verification/validation should be carried out in these domains individually, as well as cross domains. The huge complexity of SoCs (tens of billion transistors are involved), as well as aggressive time-to-market, also contribute to even more growth of verification/validation efforts.

When it comes to security verification and validation, not only all of the above-mentioned challenges are still in the picture, but the problem becomes even more challenging due to several reasons. The first reason is that security is a generic term and it is unclear how to achieve a secure design. There is no security specification or security verification plan to check the implementation against it. As we discussed in Chap. 1, there are several security vulnerabilities, including information leakage, side-channel leakage, access control violations, malicious functionality, etc., that a security verification engineer should check. Checking the implementation against those vulnerabilities requires a vast knowledge about security attacks and their targets. However, there is a lack of understanding about security issues by the designers. Designers often make decisions based on performance, constraints on design budgets, and testability. They may be unaware of the effect of their decision on creating potential security threats. On the other hand, protecting the design against one security vulnerability may make it vulnerable to the other one. For example, protecting a design against information leakage may create side-channel

© Springer Nature Switzerland AG 2020
F. Farahmandi et al., *System-on-Chip Security*,
https://doi.org/10.1007/978-3-030-30596-3_2

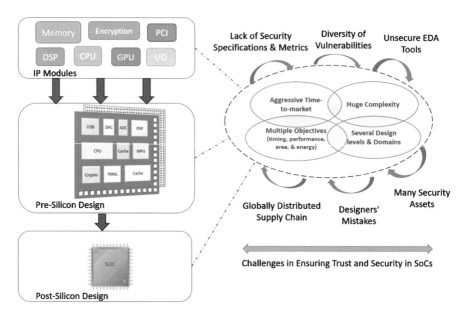

Fig. 2.1 Challenges in security verification and validation of SoC designs on the top of challenges for conventional functional verification approaches

leakage that can be exploited by an attacker to retrieve sensitive information. As of now, there is no comprehensive guideline that designers can follow and create secure hardware components.

In addition to the above-mentioned challenges, there are other factors and challenges that should be considered for verifying the security of an SoC. Figure 2.1 summarizes the challenges in security verification and validation. We briefly describe each of them in the following categories:

- **Diversity of Assets:** An SoC contains several assets that should be protected from an adversary. Secret keys are embedded in the device to perform on-chip encryption, decryption, and hashing algorithms. Unauthorized access to these codes will create confidentiality issues. Moreover, there are developer keys and configurations bits that configure/control critical operations in the design. Breaking the integrity of those will bypass the security of the system. For example, most of the encryption algorithms require a random number (nonce) to operate. An attacker should not be able to change the configuration of True Random Number Generator (TRNG) unit to produce a constant number (several options such as generating a constant number exist in the implementation of a TRNG for testing purposes) as the nonce or weaken the randomness of the generated number to break the security of the encryption unit. Chip manufacturing codes such as Original Equipment Manufacturer (OCM) and Original Component Manufacturer (OEM) keys also exist in SoCs. Compromising them would result in counterfeiting. Moreover, there may be on-device sensitive data about the user credentials that unauthorized access to them could result in breaking the privacy. Additionally, a design house may lock the gate-level netlist (obfuscate the original functionality using the placement of secret keys) of its design before sending it to the foundry to avoid overproduction. The correct key of the locked design will be placed in the chip in the design house after manufacturing to create the correct functionality. Access to the obfuscation key will lead to revenue loss for the design house.

To perform security validation, we need to precisely identify the assets of an SoC and analyze their attributes. It is a challenging task due to two main reasons: (1) as we mentioned, there are many assets in an SoC, and (2) the assets cannot be considered as static values. New assets and critical data will be introduced when the original assets are propagating to different components and affect other variables during runtime. Moreover, we need to identify threat models for each of those assets and define security rules about how they should be securely transmitted through various communication channels.

- **Lack of Security Metrics:** There is no set of comprehensive metrics so the security of each design can be quantitatively measured. Security metrics are needed to define a way to guide verification efforts and create closures (similar to branch coverage, reach ability analysis, statement coverage, etc. that are used in functional verification) for validation activities. For example, to be able to check side-channel vulnerabilities, metrics should be defined to measure unbalanced execution paths [48] that create exploitable timing and power signatures in hardware designs. Moreover, the effect of different keys on the power consumption of encryption components should be measured to avoid leaky implementations [85].

  If we consider hardware Trojans and malicious functionality, the diversity of hardware Trojans with different objectives, trigger, and payloads makes it very difficult to construct a fault model to detect them systematically. A smart adversary usually introduces Trojans with stealthy behavior and hides them in hard-to-detect areas of the design to avoid detection using conventional validation efforts. Similarly, an asset can be leaked via various forms to observable points. It is complicated to construct a model for information leakage. Therefore, a set of metrics is needed to measure the vulnerability of a design against different threat models and attack surfaces to be able to automatically identify the source of vulnerabilities and perform security verification.

- **Unintentional Vulnerabilities:** Not all of the vulnerabilities are created in the system by adversaries. Some vulnerabilities are introduced in the design by designers' mistakes, by electronic design automation (EDA) tools, or from design-for-test (DFT) and design-for-debug (DFD) infrastructures. Recently, it has been shown that speculative execution units that exist in modern processors to enhance the performance allow programs to steal secret stored in the memory of other programs [65, 66]. When a control flow of a program is dependent on an uncached value which resides in external memory, the CPU should be idle for several hundred clock cycles until the value becomes known. Rather than wasting several clock cycles, micro-architectural units (e.g., branch predictors) in modern CPUs try to execute the program on a guessed path speculatively. If the guessed path is wrong, the speculative execution will be discarded by reverting intermediate checkpoints. Otherwise, the result of the speculative execution will be committed, which causes a significant gain in performance. However, wrong execution paths create fingerprints in caches which can be exploitable using cache timing attacks [69, 103]. From the security standpoint, speculative executions may lead to an unauthorized access to assets in the design and reveal their values using side-channel analysis.

It has been shown that EDA tools such as synthesis tools unintentionally create vulnerabilities in the design [81, 82]. Synthesis tools create additional don't care states and transitions in the controller designs for optimization purposes. If don't care states are connected to the protected states of the design, an attacker can exploit fault injection attacks to reach to those states and subsequently reach to the protected states of the design. Therefore, they can bypass the security mechanisms of the circuit and change the control flow of the design. The choice of the encoding of finite state machines (FSMs) has an effect on the connectivity of don't care states to original states of the FSM. Some encoding styles create more vulnerabilities that others [82].

Considering the complexity of today's SoCs, DFD and DFT infrastructures are required for post-silicon debug and validation efforts. However, there is an inherent conflict between increasing observability and trust. Although designing effective debug infrastructures can drastically reduce the post-silicon validation and debug efforts by expanding the observability and controllability, the extra observability/controllability generated by DFD and DFT infrastructures can facilitate integrity and confidentiality issues such as trace buffer attack [49, 50] and scan-based side-channel attacks [64].

Note that all of the vulnerabilities mentioned above have been introduced unintentionally due to the fact that the current design and valuation flows are not security-aware. Therefore, existing design and verification procedures and flows should be revised to consider security.

- **Globally Distributed Supply Chain:** SoC supply chain is globally distributed over the globe. So many countries and companies around the world are involved in different stages of design, fabrication, and testing of an SoC. An SoC designer may integrate so many IPs gathered from third-party vendors in the final design to be able to decrease the cost of the design and meet time-to-market requirements. However, the third-party IPs may come with deliberate malicious functionality which targets the change of the correct behavior of the design, denial of service, causing information leakage, etc. Malicious functionality is hidden in a way to escape traditional validation efforts.

  An SoC design may be sent to outsider venues for different purposes such as power optimization, clock-tree insertion, and DFT and DFD insertion. The untested venue has full observability to the gate-level/layout of the design and can create several vulnerabilities, such as reverse engineering of the design and steal it. They can also introduce hard-to-detect malicious functionality in the design. An untrustworthy foundry could add similar threats during the fabrication of the design. All in all, the globally distributed supply chain of SoCs creates several unique vulnerabilities for SoCs that should be checked using security validation/verification techniques.

It is of paramount importance to detect and address security vulnerabilities during design and verification time. If the vulnerabilities reach the post-silicon stage, there would be limited flexibility (almost none) in changing or fixing them. Moreover, the cost of fixing the design is significantly higher as we advance through the later stages of the design (rule of ten) [80]. Furthermore, vulnerabilities that reach the manufacturing stage will cause revenue loss. Therefore, it is essential to develop efficient security validation and verification approaches to ensure the security and trustworthiness of hardware designs. In this chapter, we review the existing security validation methods for SoCs at different design stages. There has been plenty of research on trust validation at IP-level, as well as during pre-silicon and post-silicon validation. These methods focus on simulation-based approaches, side-channel analysis, and formal approaches as shown in Fig. 2.2. Simulation-based techniques focus on generating tests and utilizing simulation

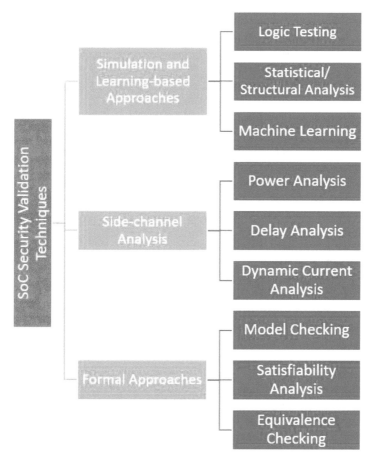

Fig. 2.2 Hardware trust verification can be categorized into three major directions: (1) simulation-based approaches, (2) side-channel analysis, and (3) formal approaches

traces to detect security vulnerabilities. Side-channel analysis approaches focus on analyzing physical characteristics of design as well as its side-channel signatures (such as current [51], leakage power [11, 58], path delay [54], electromagnetic waves [44], etc.) to detect exploitable side-channel leakage that facilities attacks on an SoC. Formal methods evaluate the security and trust of a design using mathematical models and representations. The remainder of this chapter reviews these approaches in detail.

2.2   Simulation-Based Trust Validation Approaches

Simulation-based approaches aim on generating tests to activate malicious modifications (hardware Trojans) and propagate the payload of the Trojan to primary outputs to check with the golden circuit. The difficulty of logic testing is to generate efficient tests to activate and propagate the effect of Trojans, which are stealthy enough to hide through the traditional manufacturing testing.

A major problem with the design validation is that we do not know whether a Trojan exists, and if it does, how to quickly detect and fix it. We can always keep on generating random tests, in the hope of activating the Trojan; however, random test generation is not effective for detecting stealthy Trojans.

Several approaches are focused on generation of guided test vectors to activate hardware Trojans. Traditional test generation techniques may not be beneficial as Trojans are designed in a way that they will be activated under very rare sequences of the inputs. In this section, we review simulation-based validation approaches including rare node activation, redundant circuit detection, N-detect ATPG, and code coverage techniques.

2.2.1   Logic Testing

Test generation is extremely important for both functional and trust validation of integrated circuits. A good set of tests can activate/detect vulnerabilities, facilitate finding the source of them, and help verification engineers to effectively address them. Test generation techniques can be classified into three different categories: random, constrained-random [1], and directed [18, 19, 21, 71, 79, 87]. Random test generators are used to activate unknown errors; however, relying on random tests to activate a target is inefficient when designs are large and complex. Constrained-random test generation tries to guide random test generator towards finding test vectors that may activate a set of important/interesting scenarios. The probabilistic nature of these constraints may lead to situations where the generated tests are inefficient. Moreover, constraint generation is not possible when we do not have any knowledge about the potential errors/vulnerabilities. A directed test generator, on the other hand, generates one test to target a specific scenario [20, 32, 67, 86, 88].

Clearly, less effort is needed to reach the same coverage goal using directed tests compared to random or constrained-random tests. However, existing directed test generation methods require a fault list, the set of targeted vulnerabilities, or desired functional behaviors that need to be activated [67]. Directed test generation approaches are either based on formal methods (e.g., SAT solvers and model checkers) or based on Concolic testing (combination of concrete simulation and symbolic execution techniques) [3, 5, 70, 89, 98]. These approaches cannot generate directed tests when the target (faulty scenario) is unknown.

In a case study [104], code coverage analysis and automatic test pattern generation (ATPG) are employed to identify Trojan-inserted circuits from Trojan-free circuits. The presented method utilizes test vectors to perform formal verification and code coverage analysis in the first step. If this step cannot detect existence of the hardware Trojan, some rules are checked to find unused and redundant circuits. In the next step, the ATPG tool is used to find some patterns to activate the redundant/dormant Trojans. Code coverage analysis is done over RTL models to make sure that there are no hard-to-activate events or corner-case scenarios in the design, which may serve as a backdoor and leak secret information [6, 104]. However, Trojans may exist in a design that have 100% code coverage.

Chakraborty et al. proposed MERO [16] which excites the rare nodes multiple times in order to increase the likelihood of Trojan activation. Generating such directed tests is extremely difficult given the stealthiness of activation condition. Besides, this technique is only applicable to gate-level designs and does not guarantee whether the generated tests can activate the Trojans. Usually complete coverage is required to detect Trojans [104]. Saha et al. [94] extended MERO by proposing an approach using genetic algorithm to generate tests to activate. Cruz et al. have proposed a test generation technique that combines the strength of model checking and ATPG for fast test generation [24]. Their approach partitions the design based on the scan chain. Constraints are generated for non-scan elements using model checking. These constraints as well as the scan elements are then given to ATPG for test generation. This approach is suitable only for partial scan-chain inserted designs. However, none of the existing techniques is scalable to activate and detect hidden Trojans. Moreover, logic testing would be beneficial when it uses efficient test vectors that can satisfy the Trojan triggering conditions, as well as propagate the activation effect to the observable points such as primary outputs. In other words, the test needs to reveal the existence of the malicious functionality. These kinds of tests are hard to create since trigger conditions may be satisfied after long hours of operation and they are usually designed with low probability. As a result, traditional use of existing test generation tools can be impractical to produce patterns to activate trigger conditions.

2.2.2 Statistical Methods

Statistical detection of malicious functionality relies on identifying the potentially untrustworthy circuit from the safe version using properties of known vulnerabilities. For example, FANCI [102] tries to detect hardware Trojans by marking the gates that weakly influence output signals as suspicious. FANCI uses approximate truth table for each signal to infer its effect on the outputs. However, FANCI has a high false positive rate. False positives are situations that a test marks a design as untrustworthy when in fact it is safe. A similar method named VeriTrust marks redundant logic gates as suspicious [106]. Initially, all gates that are not covered during verification phase are considered as suspicious nodes, and further analysis is carried out to confirm redundancy. FANCI and VeriTrust can detect only Trojans with always on or combinational type triggers (a trigger that depends only on current inputs). They cannot detect sequential Trojans, which are exploited by DeTrust benchmarks [105]. Hicks et al. proposed an approach for defeating Trojan based on unused circuit detection [46]. This method relies on the assumption that Trojan circuits will reside on unused portion of the circuit. However, their algorithm failed to detect Trojans that do not rely on unused circuits [100].

A score-based classification method for detecting Trojan is discussed in [84]. The classification features are based on properties found from Trojans in Trust-HUB benchmarks [101]. Scores are given to nets for each of the matching features. Nets with score above a threshold are marked as Trojan nodes. Unfortunately, these features are too specific to Trust-HUB benchmarks and thus cannot be used as a generic detection method. A recent approach proposed by Salmani et al. [95] uses SCOAP[1] controllability and observability values to detect and isolate Trojan nodes. Controllability is defined as the number of primary inputs that must be manipulated to control a signal to a particular logic value. Observability is the number of primary input manipulations which is required to make a signal observable at the primary outputs. This method works using the assumption that Trojan nodes will have higher controllability/observability values to avoid detection. However, this approach will result in false positives in designs with partial scan chains. Benign signals that are not part of the scan chain will also have controllability/observability values similar to Trojans. A Trojan clustering approach based on signal correlation is proposed in [15]. However, this method is suitable for gate-level designs, and cannot be extended to RTL models for early detection.

---

[1] SCOAP: Sandia Controllability/Observability Analysis Program [38].

2.2.3   *Machine Learning Approaches*

Several approaches have been proposed to use machine learning to detect security vulnerabilities in the system. Hasegawa et al. [42] proposed a Trojan detection technique based on static support-vector-machine-based (SVM-based) classification of gate-level netlists. This method extracts five features for any potentially suspicious candidate S in a gate-level netlist to differentiate a Trojan-inserted netlist from a safe one. The five features include the number of fan-ins of gate S (up to two levels), the number of logic levels to the closets flip-flop input from S, the number of logic levels to the closets flip-flop output from S, the minimum cone of influence from any primary input to s, and minimum propagation cone from S to any primary outputs. The features are constructed based on the stealthy behavior of hardware Trojans (as they are hardly activated and their effect may be masked in most of the cases). An SVM classifier is trained on these features to detect unknown Trojans. A similar runtime approach has been proposed for many-core platform [61]. These approaches can achieve high accuracy ($\sim$80%) in detecting Trojans. However, they also have a high false positive rate and mark many benign components as suspicious. The accuracy of such approaches has been improved by better feature selection and use of other machine learning models (e.g., neural networks) [43, 53] to reduce false positive rate.

Machine learning can build the pattern of side-channel fingerprints [55, 56] of normal circuits and any outlier will be a Trojan circuit. Most of these approaches utilize the traces of applying different tests to extract features and train the model to detect untrustworthy behavior of the design at the runtime. In these techniques, a machine is modeled and trained based parametric signature of a chip [68]. The signature is collected and compared to a trusted region in a multidimensional space. The fingerprint of Trojan-free chips is expected to fall within this region, while the fingerprint of Trojan-infested chips is expected to fall outside. However, the classification boundary is very narrow due to the uncertainty incurred by process variations. Therefore, the chance of misclassification is high.

2.3   Security Validation Using Side-Channel Analysis

Existing techniques based on side-channel analysis rely on the change of physical characteristics caused by the security vulnerability—mostly in the form of current, power, or delay [54, 68–70, 83]. If the side-channel signature of a chip is different from the golden chip over a certain threshold, a Trojan is detected. For example, when a Trojan is partially or fully activated, it will have increased switching activity compared to Trojan-free circuit. Moreover, implementation of security critical components (e.g., encryption/decryption units) should be verified to ensure that none of secret information or assets in the design can be extracted using analysis of power or delay analysis.

2.3.1   Trojan Activation Using Transient Current and Power Analysis

Side-channel approaches for detecting Trojans depend on measuring and monitoring physical parameters like transient current or power signature of an IC in order to identify sudden/unexpected changes in those parameters when the Trojan is activated. In other words, these techniques are based on the idea that the side-channel signature of a Trojan-inserted design (when the Trojan is activated) is different than the signature of a Trojan-free design. Wang et al. used this property to isolate Trojan [64]. Leakage current and transient current measurements have been widely used to detect manufacturing defects [2, 91]. These approaches are very promising to activate unknown Trojans. However, the success of these approaches is limited due to variations in processes (since Trojans are designed as small malfunctions in a covert manner and their effect on side-channel signatures is minimal). This is because the difference in side-channel signature, which is due to the Trojan, can be negligible compared to process variations.

Several approaches have been proposed to overcome process variations by maximizing the activity of rare nodes, as well as the possibility of activating Trojans. MERS utilized test generation to improve the Trojan detection sensitivity [51, 52]. Their approach selected the nodes with low transition probability as suspicious nodes. Then test vectors are applied in such a way that switching activity of these suspicious nodes become much higher than other nodes, increasing side-channel emission. However, these methods also require Trojan-free golden reference models. As side-channel analysis is carried out after fabrication, the chip may require respins if Trojan is detected. The authors in [68] relieve this limitation by using fingerprints from process control monitors. Thus, methods that can detect Trojan in an early design stage are highly desirable.

2.3.2   Trojan Detection Using Delay Analysis

Path-delay-based analysis methods can be used to detect subtle changes in the delay introduced by the trigger and payload gates of hardware Trojans. The main idea is that Trigger and payload nodes introduce additional capacitive load to the nodes that are connected and cause noticeable delay in the existing paths of the design, creating an observer effect [54, 96]. Specifically, path-delay tests are used to determine if an adversary has added a fanout to logic gate inputs and outputs to increase the delay of paths and modify the correct functionality of the design. These approaches are susceptible to generating false positives and false negatives. Ismari et al. used an on-chip measurement structure called a time-to-digital converter (TDC) to measure path delay with high resolutions [54]. This technique explores various process variation calibration methods (chip-to-chip and within-die) to improve false negative and false positive detection decisions. A technique called REBEL [14] has been proposed to use a delay chain to obtain timing information. Chapter 11 describes these techniques in detail.

2.3.3   Detecting Side-Channel Leakage Using Power Analysis

Weak implementation of a crypto algorithm can lead to side-channel leakage of sensitive information (e.g., encryption keys) despite the mathematical robustness of encryption algorithms. Therefore, an attacker can exploit such vulnerabilities to break encryption algorithms. Several power analysis techniques such as differential power analysis [58], correlation power analysis [11], partitioning power analysis [62], and template attacks [17] have been used to extract the secret key of encryption/decryption. These attacks are based on the fact that different key values may cause different power consumption in a leaky implementation of a crypto algorithm. Therefore, an attacker makes an assumption on the key and guess a hypothetical model for power consumption based on the key value. Next, he/she can utilize statistical analysis to compare physical power output and hypothetical power output to determine whether the guessed key is right.

These attacks pose major threats to the security community and several countermeasures such as hiding and masking [72] have been proposed to remove the dependency between the different key values (as well as intermediate values) and power consumption. However, these countermeasures introduce the additional area and which makes it infeasible to apply them resource-constraint designs. Different techniques have been presented to address the limitation of these countermeasures by performing side-channel leakage assessment to identify the leaky components of the design and apply countermeasures only to those parts of it. Side-channel assessment techniques include measuring signal-to-noise ratio (SNR) [75], test vector leakage assessment (TVLA) [7], and success rate [37]. These assessment techniques can be mostly applied to the post-silicon designs when it is too late to fix leaky components. He et al. [45] have proposed a side-channel assessment technique, which can be applied on RTL designs while there is still flexibility to make design changes. This method utilizes Kullback–Leibler (KL) divergence and success rat metrics to estimate the statistical distance between two different probability distributions of power signatures (based on different key pairs) and detect leaky blocks. The success of this method is dependent on the implementation of detailed and high-precision power consumption models for RTL designs.

2.4   Security Validation Using Formal Methods

Formal methods are promising in hardware validation as they evaluate the functionality and security of the design using mathematical models. Formal verification methods can be broadly classified into four categories: (1) satisfiability (SAT) solvers, (2) property checking using model checkers [10, 74], (3) information flow tracking, and (4) equivalence checking using decision graphs [12, 13, 22] and symbolic algebra [26]. In this section, we briefly discuss each of these methods and their applications for security validation. We focus on different formal methods to verify the design and detect undesired functions, unauthorized accesses, and information leakage [4, 24, 28, 29, 33, 39–41, 82].

2.4.1 Trust Validation Using SAT Solvers

Given a Boolean formula, the satisfiability problem relies on finding Boolean values to the formula's variables such that the formula is evaluated to true. If such an assignment does not exist, the formula is called unsatisfiable. The Boolean formula is constructed from AND, OR, and NOT operators between various variables which can be either assigned to true or false. Many of the validation and debugging problems can be mapped to satisfiability problems [8, 9, 59]. One of the applications is to check the equivalence between the specification of the circuit and its implementation using SAT solvers. Figure 2.3 shows the equivalence checking using SAT solvers. If hardware Trojans exist in the implementation, the SAT solver finds assignments to the internal variables to reveal the hidden Trojan. However, this method requires a golden model and suffers from scalability issues. The SAT solver may encounter state explosion when the design is large, and the specification and the implementation significantly differ from each other.

Several works explore the existence of Trojans in unspecified functionality [34, 35]. Therefore, the Trojan does not alter the specification of the design, and existing statistical or simulation-based methods cannot identify the Trojan-inserted design [36]. Fern et al. propose a SAT-based technique to detect Trojans, which exploits design signals in their unspecified functionality to cause malfunction or information leakage [36]. Suppose that the function "$func$" is unspecified when internal signal "s" is under condition "C." Suppose that signal s can have two possible values: v_0 and v_1. Under condition C, Eq. 2.1 should be unsatisfiable if the design is Trojan-free. Therefore, any assignment that makes Eq. 2.1 satisfiable is a trace (counter-example) to detect the covert Trojan. For every pair (s, C), one CNF formula is constructed and a SAT solver (for Boolean values) or a satisfiability module theory

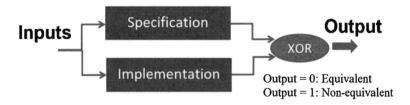

Fig. 2.3 Equivalence checking using SAT solvers [39]

(SMT) solvers can be used to find the potential threats. The success of this approach is dependent on both SAT solvers and identification of (s, C) pairs. Moreover, the approach requires manual intervention.

$$C \wedge (func(s = v_0) \oplus func(s = v_1)) \tag{2.1}$$

SAT solvers [63, 73] have been also used to automatically localize hard-to-detect bugs in arithmetic circuits. Solving SAT problem results in finding suspicious functionality. These approaches are based on either inserting logic corrector components in the implementation [99], using abstraction and refinements [57, 93] or using quantified Boolean formula [73]. The success of SAT-based approaches is dependent on the performance of SAT solvers, and they fail for large and complex circuits.

2.4.2   Security Validation Using Property (Model) Checking

Model checking is a famous technique in design verification, which checks a design for a set of given properties. To solve the model checking problem, the design and the given properties are converted to a mathematical model/language, and all of the design states are checked to see whether the given properties are satisfied. Security properties describe the expected behaviors of a secure and trustworthy design. These properties can be modeled as a collection of linear time temporal logics (LTL) [23]. A model checker either proves the correctness of a given property over all of the possible behaviors of the design or finds a counter-example when the property fails. Figure 2.4 shows the high-level overview of security verification using model checkers. The counter-example generated by model checkers can be used as a test case to activate the target scenario [20, 25, 32, 60, 76–78, 88, 90]. Since model checkers consider all of the design states, it is prone to state-space explosion issue, especially when large designs and complex properties are involved. A bounded model checking (BMC) is used to limit the design unfolding to a limited number of clock cycles. Since BMC does not check for all of the possible design states, it cannot formally prove the given property. However, BMC assumes that the designer knows the required number of clock cycles that a particular property should hold.

Model checkers can be used to ensure safety properties. An SoC designer and a third-party vendor can agree on certain security properties that should be held on the design. When the design is sent to the SoC integrator, the SoC integrator converts the design to a formal description to check the security properties using a model checker. If all of the security properties are verified, the expected security behaviors are met. Rajendran et al. have proposed a Trojan detection technique, which is based on using a bounded model checking [92]. They have considered the threat model as an attempt to corrupt the critical data such as secret keys of a cryptographic design, random numbers which are required by most of the cryptography algorithms, or stack pointer of a processor. The assumption is that these critical data should be

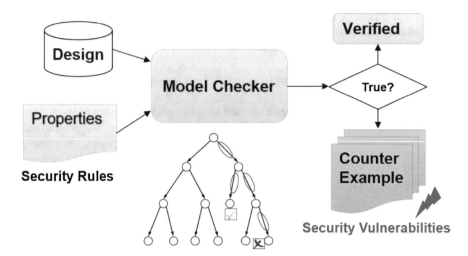

Fig. 2.4 Security verification using model checkers. Security properties describe the expected behaviors that a trustworthy design is required to follow

stored in some specific registers and accesses to these registers should be protected. In other words, the registers that contain critical data should be accessed through valid ways, and any undefined access to these registers is considered as a threat. The safe access conditions to these registers are formulated as properties (assertions), and a bounded model checker is utilized to find a counter-example when the security properties are violated.

Several security validation approaches using model checking have been proposed. J. Rajendran et al. have introduced a test generation technique for Trojan detection using BMC approach [92]. They generate a set of security properties based on access privileges to critical data registers, address tables, or stack pointer of processors. The properties are then checked against the design using BMC to find unauthorized accesses. The BMC generates a test to activate the hidden Trojan when the given property does not hold in the design. However, the strength of this approach is dependent on the completeness of the security properties, as well as the capability of the SAT solver used during BMC. A combination of information flow tracking techniques and model checking techniques has been introduced in [47]. This technique looks for confidentiality and integrity property violations. The confidentiality property requires that any secure data may not enter the unsecured domain. Conversely, the integrity property requires that anything from unsecured domain may not enter the secured domain. Use of model checking to find hardware Trojans, information leakage, and unauthorized accesses to secret and critical data is beneficial since the method does not require any golden model of the design. However, the success of this method is dependent on the SAT-engine (which may fail for large and complex designs) and precise definition of security properties, which needs prior knowledge of all safe ways to access a critical register.

2.4.2.1 Application of Equivalence Checking in Security Validation

Equivalence checking is used to formally prove that different representations of a design display the same functionality—nothing more, nothing less. Every extra, incorrect, or missing components can threaten the security of the design (the threat model is any deviation from the expected functionality). However, traditional equivalence checking methods may face a state-space explosion issue when applied on large design. There are equivalence checking methods based on symbolic algebra (Gröbner basis theory) that are successful to detect deviations from the specification for combinational circuits, especially arithmetic circuits [26, 31, 39, 97]. Using Gröbner basis theory enables formulation of the security verification problem in the algebraic domain [39]. The design specification is first converted to a specification polynomial f_{spec}. The gate-level implementation is then converted to a set of implementation polynomial F, where each gate is modeled as a polynomial. A polynomial division is then applied to reduce the specification polynomial f_{spec} over implementation polynomials F. The remainder should be zero for functionally equivalent designs. A non-zero remainder indicates that the specification differs from the implementation, and the design is not trustworthy.

The remainder not only expresses the outcome of the equivalence checking but also is beneficial in removing the Trojan. Any assignment which makes the total value of the remainder non-zero is a directed test (counter-example) that activates the Trojan. The directed test can be used to localize the source of error. Moreover, patterns and existing terms of the remainder provide valuable information to detect and correct the bug (gate misplacement and signal inversion are considered fault models) [27–30, 33]. Although this technique is promising, the complexity of this approach remains a challenge due to two factors: (1) To check whether a given polynomial is a member of a polynomial system or not, and (2) a sequence of expensive (in terms of both runtime and memory usage) polynomial manipulations, including polynomial division and multiplication, are needed.

Equivalence checking based on symbolic algebra can be used to detect malfunctions in general circuits as well. After a design is implemented and validated in pre-silicon, the synthesized gate-level netlist may go through several non-functional changes for different optimization purposes. The goal of the equivalence checking is to detect whether an adversary has inserted a hard-to-detect hardware Trojan during non-functional changes and has made undesired functional changes. For example, a design house may send their RTL design for synthesis or insertion of low-power features to a third-party vendor. Once the third-party IP comes back (after synthesis or other functionality-preserving transformations), it is crucial to ensure the trustworthiness of these IPs. In this scenario, the in-house version of the design is treated as the golden design and specification polynomials are extracted from it. Implementation polynomials are extracted from the potentially untrustworthy design. The specification and implementation polynomials are compared using polynomial manipulations and any malfunction is detected formally [33]. Similarly, equivalence checking techniques based on symbolic algebra can be used to detect the vulnerabilities in finite state machines [28].

2.5 Summary

This chapter presented challenges in security verification and validation. We also presented existing techniques for hardware security validation. It outlined prior efforts in test generation, side-channel analysis, as well as formal approaches for detecting hardware Trojans.

References

1. A. Adir, E. Almog, L. Fournier, E. Marcus, M. Rimon, M. Vinov, A. Ziv, Genesys-pro: innovations in test program generation for functional processor verification. IEEE Des. Test Comput. **21**, 84–93 (2004)
2. D. Agrawal, S. Baktir, D. Karakoyunlu, P. Rohatgi, B. Sunar, Trojan detection using IC finger-printing, in *Symposium on Security and Privacy* (2007), pp. 296–310
3. A. Ahmed, P. Mishra, QUEBS: qualifying event based search in concolic testing for validation of RTL models, in *IEEE International Conference on Computer Design (ICCD)* (2017), pp. 185–192
4. A. Ahmed, F. Farahmandi, Y. Iskander, P. Mishra, Scalable hardware Trojan activation by interleaving concrete simulation and symbolic execution, in *IEEE International Test Conference (ITC)* (2018)
5. A. Ahmed, F. Farahmandi, P. Mishra, Directed test generation using concolic testing of RTL models, in *Design Automation and Test in Europe (DATE)* (2018), pp. 1538–1543
6. M. Banga, M. Hsiao, Trusted RTL: Trojan detection methodology in pre-silicon designs, in *IEEE International Symposium on Hardware-Oriented Security and Trust (HOST)* (2010), pp. 56–59
7. G. Becker, J. Cooper, E. DeMulder, G. Goodwill, J. Jaffe, G. Kenworthy, T. Kouzminov, A. Leiserson, M. Marson, P. Rohatgi et al., Test vector leakage assessment (TVLA) methodology in practice, in *International Cryptographic Module Conference*, vol. 1001 (2013), p. 13
8. P. Behnam, B. Alizadeh, In-circuit mutation-based automatic correction of certain design errors using SAT mechanisms, in *2015 IEEE 24th Asian Test Symposium (ATS)* (IEEE, Piscataway, 2015), pp. 199–204
9. P. Behnam, B. Alizadeh, Z. Navabi, Automatic correction of certain design errors using mutation technique, in *2014 19th IEEE European Test Symposium (ETS)* (IEEE, Piscataway, 2014), pp. 1–2
10. A. Biere, A. Cimatti, E. Clarke, Y. Zhu, Symbolic model checking without BDDs, in *International Conference on Tools and Algorithms for the Construction and Analysis of Systems* (Springer, Berlin, 1999), pp. 193–207
11. E. Brier, C. Clavier, F. Olivier, Correlation power analysis with a leakage model, in *Cryptographic Hardware and Embedded Systems – CHES*, ed. by M. Joye, J.-J. Quisquater (Springer, Berlin, 2004), pp. 16–29
12. R.E. Bryant, Graph-based algorithms for Boolean function manipulation. IEEE Trans. Comput. **C-35**, 677–691 (1986)
13. R.E. Bryant, Y.-A. Chen, Verification of arithmetic circuits with binary moment diagrams, in *Proceedings of the 32nd Annual ACM/IEEE Design Automation Conference* (ACM, New York, 1995), pp. 535–541
14. M. Bushnell, V.D. Agrawal, Essentials of electronic testing for digital, memory, and mixed signal VLSI circuits, vol. 17 (Springer, New York, 2000)
15. B. Çakir, S. Malik, Hardware Trojan detection for gate-level ICS using signal correlation based clustering, in *Proceedings of the 2015 Design, Automation & Test in Europe Conference & Exhibition* (EDA Consortium, San Jose, 2015), pp. 471–476

16. R.S. Chakraborty, F. Wolf, C. Papachristou, S. Bhunia, MERO: a statistical approach for hardware Trojan detection, in *International Workshop on Cryptographic Hardware and Embedded Systems (CHES'09)* (2009), pp. 369–410
17. S. Chari, J.R. Rao, P. Rohatgi, Template attacks, in *Cryptographic Hardware and Embedded Systems – CHES*, ed. by B.S. Kaliski, Ç.K. Koç, C. Paar (Springer, Berlin, 2003), pp. 13–28
18. M. Chen, P. Mishra, Functional test generation using efficient property clustering and learning techniques. IEEE Trans. Comput. Aided Des. Integr. Circuits Syst. **29**(3), 396–404 (2010)
19. M. Chen, P. Mishra, Property learning techniques for efficient generation of directed tests. IEEE Trans. Comput. **60**(6), 852–864 (2011)
20. M. Chen, P. Mishra, Decision ordering based property decomposition for functional test generation, in *Design Automation and Test in Europe (DATE)* (2011), pp. 167–172
21. M. Chen, X. Qin, P. Mishra, Learning-oriented property decomposition for automated generation of directed tests. J. Electron. Test. **30**(3), 287–306 (2014)
22. M.J. Ciesielski, P. Kalla, Z. Zheng, B. Rouzeyre, Taylor expansion diagrams: a compact, canonical representation with applications to symbolic verification, in *Proceedings 2002 Design, Automation and Test in Europe Conference and Exhibition* (IEEE, Piscataway, 2002), pp. 285–289
23. E.M. Clarke, E.A. Emerson, and A.P. Sistla, Automatic verification of finite-state concurrent systems using temporal logic specifications. ACM Trans. Program. Lang. Syst. **8**(2), 244–263 (1986)
24. J. Cruz, F. Farahmandi, A. Ahmed, P. Mishra, Hardware Trojan detection using ATPG and model checking, in *2018 31st International Conference on VLSI Design and 2018 17th International Conference on Embedded Systems (VLSID)* (IEEE, Piscataway, 2018), pp. 91–96
25. N. Dang, A. Roychoudhury, T. Mitra, P. Mishra, Generating test programs to cover pipeline interactions, in *ACM/IEEE Design Automation Conference (DAC)* (2009), pp. 142–147
26. F. Farahmandi, B. Alizadeh, Groebner basis based formal verification of large arithmetic circuits using Gaussian elimination and cone-based polynomial extraction. Microprocess. Microsyst. **39**(2), 83–96 (2015)
27. F. Farahmandi, P. Mishra, Automated test generation for debugging arithmetic circuits, in *2016 Design, Automation & Test in Europe Conference & Exhibition (DATE)* (IEEE, Piscataway, 2016), pp. 1351–1356
28. F. Farahmandi, P. Mishra, FSM anomaly detection using formal analysis, in *IEEE International Conference on Computer Design (ICCD)* (2017), pp. 313–320
29. F. Farahmandi, P. Mishra, Automated debugging of arithmetic circuits using incremental Gröbner basis reduction, in *IEEE International Conference on Computer Design (ICCD)* (2017), pp. 193–200
30. F. Farahmandi, P. Mishra, Automated test generation for debugging multiple bugs in arithmetic circuits. IEEE Trans. Comput. **68**(2), 182–197 (2019)
31. F. Farahmandi, B. Alizadeh, Z. Navabi, Effective combination of algebraic techniques and decision diagrams to formally verify large arithmetic circuits, in *2014 IEEE Computer Society Annual Symposium on VLSI* (IEEE, Piscataway, 2014), pp. 338–343
32. F. Farahmandi, P. Mishra, S. Ray, Exploiting transaction level models for observability-aware post-silicon test generation, in *Design Automation and Test in Europe (DATE)* (2016), pp. 1477–1480
33. F. Farahmandi, Y. Huang, P. Mishra, Trojan localization using symbolic algebra, in *Asia and South Pacific Design Automation Conference (ASPDAC)* (2017), pp. 591–597
34. N. Fern, S. Kulkarni, K.-T.T. Cheng, Hardware Trojans hidden in RTL don't cares – automated insertion and prevention methodologies, in *2015 IEEE International Test Conference (ITC)* (IEEE, Piscataway, 2015), pp. 1–8
35. N. Fern, I. San, C.K. Koç, K.-T.T. Cheng, Hardware Trojans in incompletely specified on-chip bus systems, in *Proceedings of the 2016 Conference on Design, Automation & Test in Europe* (EDA Consortium, San Jose, 2016), pp. 527–530

36. N. Fern, I. San, K.-T.T. Cheng, Detecting hardware Trojans in unspecified functionality through solving satisfiability problems, in *2017 22nd Asia and South Pacific Design Automation Conference (ASP-DAC)* (IEEE, Piscataway, 2017), pp. 598–504

37. B. Gierlichs, K. Lemke-Rust, C. Paar, Templates vs. stochastic methods, in *Cryptographic Hardware and Embedded Systems – CHES 2006*, ed. by L. Goubin, M. Matsui (Springer, Berlin, 2006), pp. 15–29

38. L.H. Goldstein, E.L. Thigpen, SCOAP: Sandia controllability/observability analysis program, in *Proceedings of the 17th Design Automation Conference* (ACM, New York, 1980), pp. 190–196

39. X. Guo, R.G. Dutta, Y. Jin, F. Farahmandi, P. Mishra, Pre-silicon security verification and validation: a formal perspective, in *ACM/IEEE Design Automation Conference (DAC)* (2015), pp. 145:1–145:6

40. X. Guo, R.G. Dutta, P. Mishra, Y. Jin, Scalable SoC trust verification using integrated theorem proving and model checking, in *IEEE International Symposium on Hardware Oriented Security and Trust (HOST)* (2016), pp. 124–129

41. X. Guo, R.G. Dutta, P. Mishra, Y. Jin, Automatic code converter enhanced PCH framework for SoC trust verification. IEEE Trans. Very Large Scale Integr. VLSI Syst. **25**(12), 3390–3400 (2017)

42. K. Hasegawa, M. Yanagisawa, N. Togawa, Hardware Trojans classification for gate-level netlists using multi-layer neural networks, in *IEEE 23rd International Symposium on On-Line Testing and Robust System Design (IOLTS)* (2017), pp. 227–232

43. K. Hasegawa, Y. Shi, N. Togawa, Hardware Trojan detection utilizing machine learning approaches, in *Proceedings – 17th IEEE International Conference on Trust, Security and Privacy in Computing and Communications and 12th IEEE International Conference on Big Data Science and Engineering, TrustCom/BigDataSE 2018* (2018), pp. 1891–1896

44. J. He, Y. Zhao, X. Guo, Y. Jin, Hardware Trojan detection through chip-free electromagnetic side-channel statistical analysis. IEEE Trans. Very Large Scale Integr. VLSI Syst. **25**(10), 2939–2948 (2017)

45. M. He, J. Park, A. Nahiyan, A. Vassilev, Y. Jin, M. Tehranipoor, RTL-PSC: automated power side-channel leakage assessment at register-transfer level, in *IEEE VLSI Test Symposium (VTS)* (2019)

46. M. Hicks, M. Finnicum, S.T. King, M.M. Martin, J.M. Smith, Overcoming an untrusted computing base: detecting and removing malicious hardware automatically, in *2010 IEEE Symposium on Security and Privacy (SP)* (IEEE, Piscataway, 2010), pp. 159–172

47. W. Hu, A. Ardeshiricham, M.S. Gobulukoglu, X. Wang, R. Kastner, Property specific information flow analysis for hardware security verification, in *2018 IEEE/ACM International Conference on Computer-Aided Design (ICCAD), San Diego, CA* (2018), pp. 1–8

48. W. Hua, Z. Zhang, G.E. Suh, Reverse engineering convolutional neural networks through side-channel information leaks, in *55th ACM/ESDA/IEEE Design Automation Conference (DAC)* (2018), pp. 1–6

49. Y. Huang, P. Mishra, Trace buffer attack on the AES cipher. J. Hardw. Syst. Secur. **1**(1), 68–84 (2017)

50. Y. Huang, A. Chattopadhyay, P. Mishra, Trace buffer attack: security versus observability study in post-silicon debug, in *IEEE International Conference on Very Large Scale Integration (VLSI-SoC)* (2015), pp. 355–360

51. Y. Huang, S. Bhunia, P. Mishra, MERS: statistical test generation for side-channel analysis based Trojan detection, in *Proceedings of the 2016 ACM SIGSAC Conference on Computer and Communications Security* (ACM, New York, 2016), pp. 130–141

52. Y. Huang, S. Bhunia, P. Mishra, Scalable test generation for Trojan detection using side channel analysis. IEEE Trans. Inf. Forensics Secur. **13**(11), 2746–2760 (2018)

53. T. Inoue, K. Hasegawa, Y. Kobayashi, M. Yanagisawa, N. Togawa, Designing subspecies of hardware Trojans and their detection using neural network approach, in *IEEE 8th International Conference on Consumer Electronics* (2018), pp. 1–4

54. D. Ismari, J. Plusquellic, C. Lamech, S. Bhunia, F. Saqib, On detecting delay anomalies introduced by hardware Trojans, in *2016 IEEE/ACM International Conference on Computer-Aided Design (ICCAD), Austin, TX* (2016), pp. 1–7

55. Y. Jin, Y. Makris, Hardware Trojan detection using path delay fingerprint, in *IEEE International Workshop on Hardware-Oriented Security and Trust* (2008), pp. 51–57

56. Y. Jin, Y. Makris, Hardware Trojans in wireless cryptographic ICs. IEEE Des. Test Comput. **27**(1), 26–35 (2010)

57. B. Keng, A. Veneris, Path-directed abstraction and refinement for SAT-based design debugging. IEEE Trans. Comput. Aided Des. Integr. Circuits Syst. **32**(10), 1609–1622 (2013)

58. P.C. Kocher, J. Jaffe, B. Jun, Differential power analysis, in *Proceedings of the 19th Annual International Cryptology Conference on Advances in Cryptology, Series CRYPTO '99, London, UK* (Springer, London, 1999), pp. 388–397. [Online]. Available: http://dl.acm.org/citation.cfm?id=646764.703989

59. H.-M. Koo, P. Mishra, Test generation using SAT-based bounded model checking for validation of pipelined processors, in *Proceedings of the 16th ACM Great Lakes Symposium on VLSI* (ACM, New York, 2006), pp. 362–365

60. H. Koo, P. Mishra, Functional test generation using property decompositions for validation of pipelined processors, in *Design Automation and Test in Europe (DATE)* (2006), pp. 1240–1245

61. A. Kulkarni, Y. Pino, T. Mohsenin, SVM-based real-time hardware Trojan detection for many-core platform, in *17th International Symposium on Quality Electronic Design (ISQED)* (2016), pp. 362–367

62. T.-H. Le, J. Clédière, C. Canovas, B. Robisson, C. Servière, J.-L. Lacoume, A proposition for correlation power analysis enhancement, in *Cryptographic Hardware and Embedded Systems – CHES 2006*, ed. by L. Goubin, M. Matsui (Springer, Berlin, 2006), pp. 174–186

63. B. Le, H. Mangassarian, B. Keng, A. Veneris, Non-solution implications using reverse domination in a modern SAT-based debugging environment, in *Design Automation and Test in Europe (DATE)* (2012), pp. 629–634

64. J. Lee, M. Tebranipoor, J. Plusquellic, A low-cost solution for protecting IPs against scan-based side-channel attacks, in *24th IEEE VLSI Test Symposium* (2006), pp. 1–6

65. C. Li, J. Gaudiot, Online detection of spectre attacks using microarchitectural traces from performance counters, in *2018 30th International Symposium on Computer Architecture and High Performance Computing (SBAC-PAD), Lyon, France* (2018), pp. 25–28

66. M. Lipp, M. Schwarz, D. Gruss, T. Prescher, W. Haas, A. Fogh, J. Horn, S. Mangard, P. Kocher, D. Genkin, Y. Yarom, M. Hamburg, Meltdown: reading kernel memory from user space, in *27th Security Symposium (USENIX Security)* (2018), pp. 973–990

67. L. Liu, S. Vasudevan, Efficient validation input generation in RTL by hybridized source code analysis, in *Design, Automation & Test in Europe Conference & Exhibition (DATE), 2011* (IEEE, Piscataway, 2011), pp. 1–6

68. Y. Liu, K. Huang, Y. Makris, Hardware Trojan detection through golden chip-free statistical side-channel fingerprinting, in *Proceedings of the 51st Annual Design Automation Conference (DAC '14)* (ACM, New York, 2014), 6 pp. Article 155

69. Y. Lyu, P. Mishra, A survey of side channel attacks on caches and countermeasures. J. Hardw. Syst. Secur. **2**, 33–50 (2018)

70. Y. Lyu, P. Mishra, Efficient test generation for Trojan detection using side channel analysis, in *Design Automation and Test in Europe (DATE)* (2019)

71. Y. Lyu, X. Qin, M. Chen, P. Mishra, Directed test generation for validation of cache coherence protocols. IEEE Trans. Comput. Aided Des. Integr. Circuits Syst. **38**, 163–176 (2018)

72. S. Mangard, E. Oswald, T. Popp, *Power Analysis Attacks: Revealing the Secrets of Smart Cards*. Advances in Information Security (Springer, New York, 2007)

73. H. Mangassarian, A. Veneris, S. Safarpour, M. Benedetti, D. Smith, A performance-driven QBF-based iterative logic array representation with applications to verification, debug and test, in *2007 IEEE/ACM International Conference on Computer-Aided Design* (IEEE, Piscataway, 2007), pp. 240–245

74. K.L. McMillan, *Model Checking* (Wiley, New York, 2003)
75. T.S. Messerges, E.A. Dabbish, R.H. Sloan, Examining smart-card security under the threat of power analysis attacks. IEEE Trans. Comput. **51**(5), 541–552 (2002). [Online] Available: https://doi.org/10.1109/TC.2002.1004593
76. P. Mishra, M. Chen, Efficient techniques for directed test generation using incremental satisfiability, in *International Conference on VLSI Design* (2009), pp. 65–70
77. P. Mishra, N. Dutt, Graph-based functional test program generation for pipelined processors, in *Design Automation and Test in Europe (DATE)* (2004), pp. 182–187
78. P. Mishra, N. Dutt, Functional coverage driven test generation for validation of pipelined processors, in *Design Automation and Test in Europe (DATE)* (2005), pp. 678–683
79. P. Mishra, N. Dutt, Specification-driven directed test generation for validation of pipelined processors. ACM Trans. Des. Autom. Electron. Syst. **13**(2), 36 pp., article 42 (2008)
80. P. Mishra, S. Bhunia, M. Tehranipoor (eds.), *Hardware IP Security and Trust* (Springer, Cham, 2017). ISBN: 978-3-319-49024-3
81. A. Nahiyan, K. Xiao, K. Yang, Y. Jin, D. Forte, M. Tehranipoor, AVFSM: a framework for identifying and mitigating vulnerabilities in FSMs, in *2016 53nd ACM/EDAC/IEEE Design Automation Conference (DAC)* (IEEE, Piscataway, 2016), pp. 1–6
82. A. Nahiyan, F. Farahmandi, P. Mishra, D. Forte, M. Tehranipoor, Security-aware FSM design flow for identifying and mitigating vulnerabilities to fault attacks. IEEE Trans. Comput. Aided Des. Integr. Circuits Syst. **38**, 1003–1016 (2018)
83. S. Narasimhan, D. Du, R.S. Chakraborty, S. Paul, F.G. Wolff, C.A. Papachristou, K. Roy, S. Bhunia, Hardware Trojan detection by multiple-parameter side-channel analysis. IEEE Trans. Comput. **62**(11), 2183–2195 (2013)
84. M. Oya, Y. Shi, M. Yanagisawa, N. Togawa, A score-based classification method for identifying hardware-Trojans at gate-level netlists, in *Proceedings of the 2015 Design, Automation & Test in Europe Conference & Exhibition* (2015), pp. 465–470
85. J. Park, A. Nahiyan, A. Vassilev, Y. Jin, M. Tehranipoor, RTL-PSC: automated power side-channel leakage assessment at register-transfer level. arXiv preprint arXiv:1901.05909 (2019)
86. S. Proch, P. Mishra, Test generation for hybrid systems using clustering and learning techniques, in *International Conference on VLSI Design* (2016), pp. 589–590
87. X. Qin, P. Mishra, Directed test generation for validation of multicore architectures. ACM Trans. Des. Autom. Electron. Syst. **17**(3), article 24, 21 pp. (2012)
88. X. Qin, P. Mishra, Automated generation of directed tests for transition coverage in cache coherence protocols, in *Design Automation and Test in Europe (DATE)* (2012)
89. X. Qin, P. Mishra, Scalable test generation by interleaving concrete and symbolic execution, in *International Conference on VLSI Design* (2014), pp. 104–109
90. X. Qin, M. Chen, P. Mishra, Synchronized generation of directed tests using satisfiability solving, in *International Conference on VLSI Design* (2010), pp. 351–356
91. R. Rad, J. Plusquellic, M. Tehranipoor, Sensitivity analysis to hardware Trojans using power supply transient signals, in *Workshop on Hardware-Oriented Security and Trust* (2008), pp. 3–7
92. J. Rajendran, V. Vedula, R. Karri, Detecting malicious modifications of data in third-party intellectual property cores, in *Proceedings of the 52nd Annual Design Automation Conference* (ACM, New York, 2015), pp. 1–6
93. S. Safarpour, A. Veneris, Abstraction and refinement techniques in automated design debugging, in *Seventh International Workshop on Microprocessor Test and Verification (MTV'06)* (IEEE, Piscataway, 2006), pp. 88–93
94. S. Saha, R. Chakraborty, S. Nuthakki, Anshul, D. Mukhopadhyay, Improved test pattern generation for hardware Trojan detection using genetic algorithm and Boolean satisfiability, in *Cryptographic Hardware and Embedded Systems (CHES)* (2015), pp. 577–596
95. H. Salmani, COTD: reference-free hardware Trojan detection and recovery based on controllability and observability in gate-level netlist. IEEE Trans. Inf. Forensics Secur. **12**(2), 338–350 (2017)

96. F. Saqib, D. Ismari, C. Lamech, J. Plusquellic, Within-die delay variation measurement and power transient analysis using REBEL. IEEE Trans. Very Large Scale Integr. VLSI Syst. **23**(4), 776–780 (2015)

97. A. Sayed-Ahmed, D. Gro, M. Soeken, R. Drechsler et al., Formal verification of integer multipliers by combining gröbner basis with logic reduction, in *2016 Design, Automation & Test in Europe Conference & Exhibition (DATE)* (IEEE, Piscataway, 2016), pp. 1048–1053

98. K. Sen, G. Agha, CUTE and jCUTE: concolic unit testing and explicit path model-checking tools, in *International Conference on Computer Aided Verification* (Springer, Berlin, 2006), pp. 419–423

99. A. Smith, A. Veneris, M.F. Ali, A. Viglas, Fault diagnosis and logic debugging using Boolean satisfiability. IEEE Trans. Comput. Aided Des. Integr. Circuits Syst. **24**(10), 1606–1621 (2005)

100. C. Sturton, M. Hicks, D. Wagner, S.T. King, Defeating UCI: building stealthy and malicious hardware, in *2011 IEEE Symposium on Security and Privacy (SP)* (IEEE, Piscataway, 2011), pp. 64–77

101. Trust-HUB, https://www.trust-hub.org/

102. A. Waksman et al., FANCI: identification of stealthy malicious logic using Boolean functional analysis, in *CCS* (2013), pp. 697–708

103. Y. Yarom, K. Falkner, FLUSH+ RELOAD: a high resolution, low noise, L3 cache side-channel attack, in *USENIX Security Symposium*, vol. 1 (2014), pp. 22–25

104. X. Zhang, M. Tehranipoor, Case study: detecting hardware Trojans in third-party digital IP cores, in *2011 IEEE International Symposium on Hardware-Oriented Security and Trust (HOST)* (IEEE, Piscataway, 2011), pp. 67–70

105. J. Zhang, F. Yuan, Q. Xu, DeTrust: defeating hardware trust verification with stealthy implicitly-triggered hardware Trojans, in *Proceedings of the 2014 ACM SIGSAC Conference on Computer and Communications Security* (ACM, New York, 2014), pp. 153–166

106. J. Zhang, F. Yuan, L. Wei, Y. Liu, Q. Xu, VeriTrust: verification for hardware trust. IEEE Trans. Comput. Aided Des. Integr. Circuits Syst. **34**(7), 1148–1161 (2015)

Chapter 3
SoC Trust Metrics and Benchmarks

3.1 Threats to IP Trustworthiness

A typical SoC design flow is shown in Fig. 3.1. Design specification by the SoC integrator is generally the first step. The SoC integrator then identifies a list of IPs necessary to implement the given specification. These IP cores are either developed in-house or purchased from 3PIP vendors. These 3PIP cores can be procured from the vendors in one of the following three ways:

- Soft IP cores are delivered as synthesizable register transfer level (RTL) hardware description language (HDL).
- Hard IP cores are delivered as GDSII representations of a fully placed and routed core design.
- Firm IP cores are optimized in structure and topology for performance and area, possibly using a generic library.

After developing/procuring all the necessary soft IPs, the SoC design house integrates them to generate the RTL specification of the whole system. SoC integrator then synthesizes the RTL description into a gate-level netlist based on the logic cells and I/Os of a target technology library, then they may integrate gate-level IP cores from a vendor into this netlist. They also add design-for-test (DFT) structures to improve the design's testability. The next step is to translate the gate-level netlist into a physical layout based on logic cells and I/O geometries. It is also possible to import IP cores from vendors in GDSII layout file format. After performing static timing analysis (STA) and power closure, developers generate the final layout in GDSII format and send it out for fabrication.

Today's advanced semiconductor technology requires prohibitive investment for each stage of the SoC development procedure. As a result, most semiconductor companies cannot afford maintaining such a long supply chain from design to packaging. In order to lower R&D cost and speed up the development cycle, the SoC design houses typically outsource fabrication to a third-party foundry, purchase

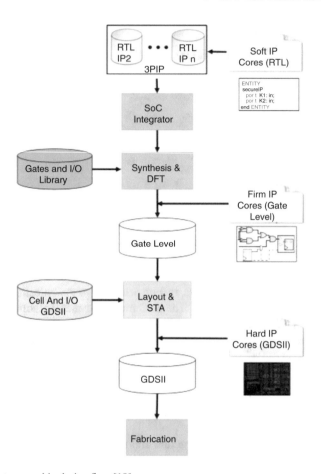

Fig. 3.1 System-on-chip design flow [19]

third-party intellectual property (IP) cores, and/or use electronic design automation (EDA) tools from third-party vendors. The use of untrusted (and potentially malicious) third parties increases the security concerns. Thus, the supply chain is now considered susceptible to various attacks, such as hardware Trojan insertion, reverse engineering, IP piracy, IC tampering, IC cloning, IC overproduction, and so forth. Among these, hardware Trojans are arguably one of the biggest concern and have garnered considerable attention. Trojans can be inserted in SoCs at the RTL, at the gate level during synthesis and DFT insertion, at the layout level during placement and routing, or during IC manufacturing. An attacker can also insert a Trojan through IP cores provided by external vendors. Designers must verify the trustworthiness of IP cores to ensure that they perform as intended, nothing more and nothing less.

3.2 IP Trust Validation

IP trust validation focuses on verifying that an IP does not perform any malicious function, i.e., an IP does not contain any Trojan. Existing IP trust validation techniques can be broadly classified into code/structural analysis, functional verification, logic testing, formal verification, and runtime validation.

Code Coverage Analysis Code coverage is defined as the percentage of lines of code that has been executed during functional verification of the design. This metric gives a quantitative measure of the completeness of the functional simulation of the design. In [11], the authors have proposed a technique named unused circuit identification (UCI) to find the lines of RTL code that have not been executed during simulation. These unused lines of codes can be considered to be part of a malicious circuit. In [3], the authors have proposed similar code coverage analysis in combination with hardware assertion checker to identify malicious circuitry in a 3PIP. However, these techniques do not guarantee the trustworthiness of a 3PIP. The authors in [26] have demonstrated that hardware Trojans can be designed to defeat UCI technique. This type of Trojans derives their triggering circuits from less likely events to evade detection from code coverage analysis.

Formal Verification Formal methods such as symbolic execution, model checking, and information flow have been traditionally applied to software systems for finding security bugs and improving test coverage. Formal verification has also shown to be effective in verifying the trustworthiness of 3PIP [9, 15]. These approaches are based on the concept of proof-carrying code (PCC) to formally validate the security-related properties of an IP. In these proposed approaches, an SoC integrator provides a set of security properties in addition to the standard functional specification to the IP vendor. A formal proof of these properties alongside with the hardware IP is then provided by the third-party vendor. SoC integrator then validates the proof by using the PCC. Any malicious modification of the IP would violate this proof indicating the presence of hardware Trojan. However, these approaches cannot ensure complete trust in an IP because the third-party vendor crafts the formal proof of these security-related properties [2].

Structural Analysis Structural analysis employs quantitative metrics to mark signals or gates with low activation probability as suspicious. In [23], the authors have presented a metric named "Statement Hardness" to evaluate the difficulty of executing a statement in the RTL code. Areas in a circuit with large value of "Statement Hardness" are more vulnerable to Trojan insertion. At gate level, an attacker would most likely target hard-to-detect areas of the gate-level netlist to insert Trojan. Inserting a Trojan in hard-to-detect areas would reduce the probability to trigger the Trojan and thereby, reduce the probability of being detected during verification and validation testing. In [29], the authors have proposed metrics to evaluate hard-to-detect areas in the gate-level netlist. The limitations of code/structural analysis techniques are that they do not guarantee Trojan detection, and manual post-processing is required to analyze suspicious signals or gates and determine if they are a part of a Trojan.

Logic Testing Logic testing aims to activate Trojans by applying test vectors and comparing the responses with the correct results. While at first glance this is similar in spirit to manufacturing tests for detecting manufacturing defects, conventional manufacturing tests using functional/structural/random patterns perform poorly to reliably detect hardware Trojans [4]. Intelligent adversaries can design Trojans that are activated under very rare conditions, so they can go undetected under structural and functional tests during the manufacturing test process. In [12, 13, 16] the authors have developed a test pattern generation methods to trigger such rarely activated nets and improve the possibility of observing Trojan's effects from primary outputs. However, this technique does not guarantee to trigger the Trojan and it is infeasible to apply this technique in an industrial scale design.

Functional Analysis Functional analysis applies random input patterns and performs functional simulation of the IP to find suspicious regions of the IP which have similar characteristics of a hardware Trojan. The basic difference between functional analysis and logic testing is that logic testing aims to apply specific patterns to activate a Trojan, whereas functional analysis applies random patterns and these patterns are not directed to trigger the Trojan. The authors in [30] have proposed a technique named functional analysis for nearly unused circuit identification (FANCI) which flags nets having weak input-to-output dependency as suspicious. This approach is based on the observation that a hardware Trojan is triggered under very rare condition. Therefore the logic implementing the trigger circuit of a Trojan is nearly unused or dormant during normal functional operation. This approach has similar drawbacks as structural analysis.

Runtime Validation Runtime validation approaches are generally based on dual modular redundancy based approaches [5, 21]. These techniques rely on procuring IP cores of same functionality from different IP vendors. The basic assumption is that it is highly unlikely that different Trojans in different 3PIPs will produce identical wrong outputs. Therefore, by comparing the outputs of IP cores of same functionality but obtained from different vendors, one can detect any malicious activity triggered by a Trojan. The main disadvantage of this approach is that the area overhead is prohibitively high and the SoC integrator needs to purchase same functional IP from different vendors (economically infeasible).

3.3 Static Benchmarks

Research in the field of hardware IP verification has seen significant growth in the past decade. Benchmarks serve as an important tool for researchers to assess the effectiveness of their proposed methodologies by providing a standardized baseline. Traditionally, Trojans have been inserted in an ad hoc manner into pre-silicon designs. This fact prevents effective comparison among Trojan detection methods because an ad hoc Trojan can favor one detection methodology over another. Recent efforts have attempted to remedy this by offering a benchmark suite with fixed number of Trojan-inserted designs [24, 25].

Standard benchmarks to evaluate hardware Trojans and their detection have been developed, such as the ones in from Trust-HUB (http://trust-hub.org/resources/ benchmarks). In this section, we focus on the hardware Trojan benchmarks and hardware obfuscation benchmarks from Trust-HUB website. We discuss the taxonomy that researchers have used for hardware security research.

3.3.1 Hardware Trojan Taxonomy

Trust-HUB has hundreds of benchmark circuits with a hardware Trojan inserted. These Trojan benchmarks are carefully crafted so that they are guaranteed to be stealthy and remain undiscovered under conventional testing methods. It is recommended that users search for trust benchmarks based on the Trojan characteristics they are interested in. The hardware Trojan taxonomy is shown in Fig. 3.2. Trojans [8, 10, 18] can be inserted at different phases (i.e., specification, design, fabrication, testing, and packaging) of the IP design flow. From the perspective of abstraction levels, Trojans can be inserted at different levels (i.e., system, development environment, RTL, gate level, layout, or physical). Activation mechanism can be either "Always On" or "Triggered." Effects of a Trojan can differ depending on the attacker's malicious intent, which might change the functionality, degrade the performance, leak information, or cause denial of service. Locations of a Trojan can be anywhere including processor, memory, I/O, power supply, or clock grid. The physical characteristics of Trojans might also vary a lot in terms of distribution, size, type, and structure.

3.3.2 Hardware Obfuscation Taxonomy

IP protection can be based on hardware obfuscation, which makes reverse engineering an IP design infeasible for adversaries and untrusted parties with any

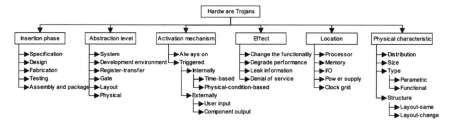

Fig. 3.2 Hardware Trojan taxonomy [25]

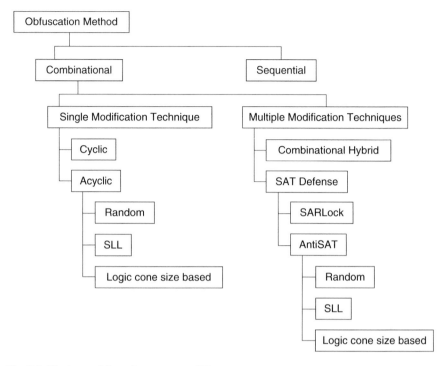

Fig. 3.3 Hardware obfuscation taxonomy [1]

reasonable amount of resources. As shown in Fig. 3.3, hardware obfuscation can be divided into two groups: combinational and sequential obfuscation. The one or more combinational obfuscation methods can be applied simultaneously on a circuit. Single techniques can be either cyclic or acyclic. The acyclic techniques applied for generating obfuscation were random insertion, secure logic locking (SLL) [32], and logic cone size [20] based. Multiple modifications can be coupling with SAT attack [31] resiliency block or using multiple obfuscation methods. The SAT resiliency block can be AntiSAT [31], SARLock [33], or any other technique. These blocks can be applied along with other combinational obfuscation blocks.

3.4 Dynamic Trojan Benchmark Generation

Existing trust benchmarks have the following major deficiencies: (1) These benchmarks only enumerate a subset of the possible hardware Trojans. There exists an inherent bias in these designs as the Trojan location and trigger conditions are static. As a result, it is possible for researchers to tune their methods (often unknowingly) to detect these Trojans. (2) A static set of benchmarks also prevent us from incorporating new types of Trojans in this rapidly evolving field, which

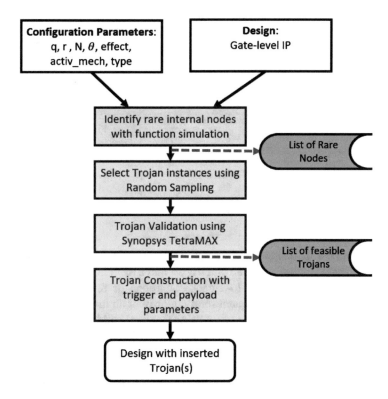

Fig. 3.4 Tool flow and integration with commercial test tool in the Trojan insertion framework [7]

keep discovering new Trojan structures. (3) A limited number of benchmarks can negatively affect supervised machine learning techniques for Trojan detection which require an expansive test set. (4) Finally, it does not allow inserting multiple Trojans in a design or inserting Trojan in a different design, e.g., a new intellectual property (IP) block.

In order to make a more robust and flexible Trust benchmark suite, we have developed a novel framework to dynamically insert various functional Trojan types into a gate-level design. The possible Trojans that can be inserted using the tool can vary in terms of the Trojan type (e.g., combinational, sequential, etc.), the number of Trojans, detection difficulty, number and rarity of trigger points, payload types, and Trojan structure. For added flexibility and forward-compatibility, we also allow users to insert a *Template Trojan* triggered from a combination of rare or non-rare nodes. Figure 3.4 provides an overview of the proposed framework. We first identify the rare internal nodes in the given netlist. Potential Trojan instances are generated using the principle of random sampling from the population of rare nodes (and non-rare if specified). Trigger conditions and payloads are verified producing a feasible Trojan list. From this list, we then randomly select and insert the Trojans according to the user's configuration options including any footprint optimization.

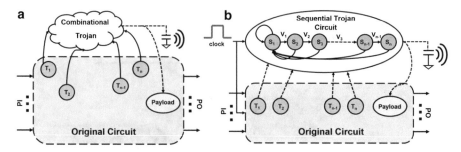

Fig. 3.5 (**a**) Generic Combinational Trojan; (**b**) Generic Sequential Trojan

3.4.1 Trojan Model

Hardware Trojans are malicious modifications of a circuit that cause undesired side-effects. Trojans structurally consist of a trigger and a payload. Typically, an attacker will insert Trojans under hard-to-trigger conditions or rare nodes of a design. Therefore, most Trojans lay dormant for a majority of an infected circuit's lifetime and subsequently evade detection when using standard validation techniques. Once the trigger or activation condition is reached, the Trojan's effect is realized through the payload gate. These payload effects can be broadly classified as functional *denial of service* (DoS) or *information leakage* [2, 28].

There are two general structures of functional Trojans: combinational and sequential. Combinational Trojans consist of only combinational logic gates. Figure 3.5a shows a generic combinational Trojan. Simultaneous activation of rare nodes T_1 to T_n leads to undesired effects. Sequential Trojans include state elements and a series of state transitions to trigger an undesired effect. In Fig. 3.5b, after the state transitions from S_1 to S_n occur, the Trojan is activated. These state transitions can be activated from trigger nodes T_1 to T_n or from just the clock signal in the case of always-on Trojans.

3.4.2 Customizable and Dynamic Trojan Insertion

Having a dynamic Trojan insertion tool is important in achieving a robust benchmark suite for hardware security. Table 3.1 describes the user configuration parameters and their effect on Trojan trigger and payload: rare node threshold θ, number of trigger nodes q, number of rare trigger nodes r, number of Trojan instances N, the effect, activation mechanism, and the Trojan type (combinational or sequential). Because we cannot include every Trojan type, we also allow the user to specify a *Template Trojan* that follows a particular format. With this input we construct the Trojan and automatically insert it into the design. Algorithm 1 shows the four major steps in the automated Trojan insertion framework shown in Fig. 3.4. Algorithm 1 takes a gate-level design D and the set of configuration parameters C from Table 3.1 and generates a Trojan-inserted design T. The remainder of this section describes these steps in detail.

Table 3.1 Configurable Trojan insertion tool parameters

Parameter	Objective
No. of trig. node (q)	Affects trigger probability, complexity
No. of rare trig. node (r)	Affects trigger probability, complexity
Rare signal threshold (θ)	Affects trigger probability and available trig. nodes
No. of Trojans (N)	Affects trigger probability, structure
Activation mechanism (always on/triggered)	Affects Trojan complexity, potential effect
Trojan effect (functional/ leakage)	Affects Trojan complexity, payload effect
Trojan structure (comb/seq/templ)	Affects Trojan complexity, payload effect

---

Algorithm 1: Dynamic Trojan insertion algorithm

    Input: Design D, set of config param $C=\{q,r,\theta,N,...\}$
    Output: Trojan inserted design T
1 $rareNode, sampleTrojPop=\{\}$
2 $hypergraph$ = constructGraph(D)
3   topologicalSort($hypergraph$)
4 $stats$ = functionalSim($hypergraph$)
5   **for** *each* node $\in$ hypergraph **do**
6 | **if** $node.signalProb \leq \theta$ **then**
7 | | $rareNode \cup node_i$
8 | **end**
9   **end**
10 trigPop = $[rareNode]^r + [node \char`\^ rareNode]^{q-r}$
11 sampleTrigPop = sample(trigPop, 10000)
12 **for** *each* $trigger \in sampleTrigPop$ **do**
13 | **if** $!validTrigger(trigger_i)$ **then**
14 | | removeTrig($trigger_i$, sampledTrigPop)
15 | **end**
16 | **else**
17 | | $payload$=findRandomPayload($hypergraph$)
18 | | **while** $!validPayload(payload)$ **do**
19 | | | $payload$=findRandomPayload($hypergraph$)
20 | | | **if** $allVisited(hypergraph)$ **then**
21 | | | | removeTrig($trigger_i$, sampledTrigPop)
22 | | | **end**
23 | | **end**
24 | | $sampleTrojPop \cup (trigger_i + payload)$
25 | **end**
26 **end**
27 T = constructTrojans(D, C)
28 **return** T

3.4.3   Identify Rare Internal Nodes

Trojans are generally difficult to detect because an adversary would likely insert a Trojan under a set of hard-to-activate internal trigger conditions. Lines 5–9 of Algorithm 1 describe how we identify rare nodes. We first construct a hypergraph by parsing the gate-level netlist. Next, the graph is sorted topologically and we employ functional simulation from a set of input patterns and compute the signal probability p for each activation level for each net. Users provide a threshold signal probability θ. All nets with a signal probability $p < \theta$ will be considered rare, and therefore potential trigger conditions. If we assume q trigger nodes with independent signal probabilities p, the resulting Trojan trigger probability becomes

$$\prod_{i=0}^{q} p_i \text{ where } p_i \in \{p\}$$

Yet, smart adversaries may try to bypass this assumption by including non-rare nodes in the Trojan trigger. This modification can evade techniques which consider only rare nodes in the activation condition. Therefore, in addition to the number of trigger nodes, we allow users to specify the number of rare trigger nodes r with $r \subseteq q$. If $r < q$, then the remaining trigger nodes will be selected from the signals with $p > \theta$ as shown in line 11 of Algorithm 1.

3.4.4   Selecting Trojans Using Random Sampling

Suppose we have identified m rare nodes with signal probability less than θ. If a Trojan can have q trigger nodes, then the total potential Trojan population for a given structure is

$$\frac{m!}{q!(m-q)!}$$

If we consider a combination of rare and non-rare trigger nodes for the Trojan, the population becomes much larger. In order to accurately model the potential Trojan population, we employ random sampling in line 11 of Algorithm 1 and select a large sample size (e.g., 10,000) potential Trojan triggers [4].

3.4.5   Trojan Validation

Inserting potential Trojans does not guarantee an adverse effect will be observable during the lifetime of the circuit. Many Trojans will be invalid or unobservable due to imposed constraints or redundant circuitry. Therefore, the inserted Trojan must be validated. In lines 12–26 of Algorithm 1, the tool verifies both the trigger condition and payload observability to ensure the Trojan inserted is a valid Trojan. From the list of potential trigger conditions, we use Synopsys TetraMAX [27] to remove any false trigger conditions by justifying each trigger condition. If the trigger condition results in a conflict, the trigger is removed from the trigger pool.

For each instance in the sampled list of feasible trigger nodes, a payload is selected randomly from the remaining nets. The criteria for payload selection is the topological order of the payload net must be greater than that of all the trigger nodes. This prevents the formation of combinational loops. Additionally, to ensure observability, stuck-at fault testing is performed for each payload. Any payload for which a stuck-at fault cannot be generated is removed from the list of feasible payloads. After combining the validated trigger condition and payload, a list of feasible Trojans exists which can be inserted into the design. If no Trojans are possible, users can increase the effort of TetraMAX or adjust the rare threshold value to include more potential Trojans.

3.4.6 Trojan Construction and Insertion

From the list of feasible Trojans, we randomly select the trigger instance and construct the Trojan from the user specification. The user can choose between functional and leakage payload effects. For functional Trojans, there are three Trojan structures a user can insert: combinational, sequential, and template. For example, Figs. 3.6 and 3.7 show a design before and after Trojan insertion. Our tool (DeTrust) for automatic Trojan insertion has a web interface as shown in Fig. 3.8.

Combinational The combinational Trojan consists of a sample structure with q trigger conditions connected together using an *AND* gate. The output of the *AND* gate is connected to an *XOR* gate. In Fig. 3.9, the left design is an example design after we generated a list of feasible Trojans. With input $\theta = 0.2$, $q = 2$, $r = 2$, $N = 2$, $type = $ comb, $effect = $ func, and $activ_mech = $ trigger the design on the right is generated.

Fig. 3.6 DeTrust template before insertion

```
and2s1 templTrig1 (.Q(trigO1), .DIN1(in[0]),
    .DIN2(in[1]));
and2s1 templTrig2 (.Q(trigO2), .DIN1(in[2]),
    .DIN2(in[3]));
dffs1 templTroj1 (.Q(troj0), .CLK(CK),
    .DIN(trigO1) );
dffs1 templTroj2 (.Q(troj1), .CLK(CK),
    .DIN(trigO2));
and2s1 templTroj3 (.Q(troj2), .DIN1(troj0),
    .DIN2(troj1));
dffs1 templTroj4 (.Q(troj3), .CLK(CK),
    .DIN(troj2));
xor2s1 templPayload (.Q(payload),
    .DIN1(original), .DIN2(troj2));
```

Fig. 3.7 DeTrust template
after insertion

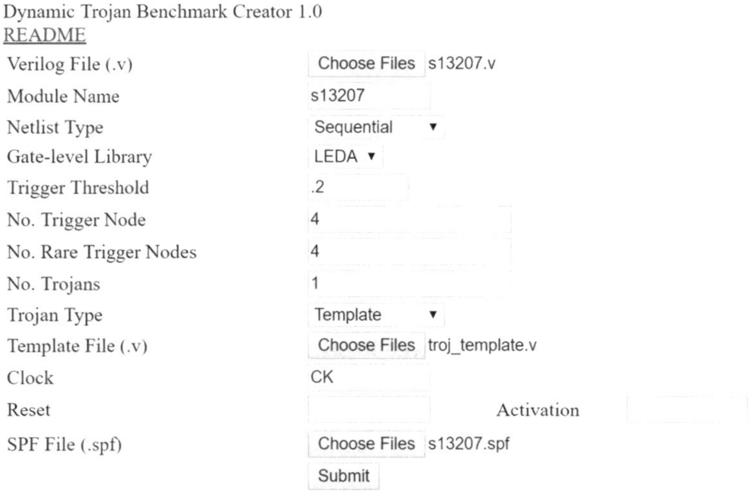

```
hi1s1 inv0 (.Q(iOut1), .DIN(n15));
and2s1 templTrig1 (.Q(trigO1), .DIN1(n17),
  .DIN2(iOut1));
and2s1 templTrig2 (.Q(trigO2), .DIN1(n12),
  .DIN2(n18));
dffs1 templTroj1  (.Q(troj0), .CLK(CK),
  .DIN(trigO1) );
dffs1 templTroj2 (.Q(troj1), .CLK(CK),
  .DIN(trigO2));
and2s1 templTroj3 (.Q(troj2), .DIN1(troj0)
  .DIN2(troj1));
dffs1 templTroj4 (.Q(troj3), .CLK(CK),
  .DIN(troj2));
xor2s1 templPayload (.Q(n26),
  .DIN1(n26_temp), .DIN2(troj2));
```

Dynamic Trojan Benchmark Creator 1.0
README

Verilog File (.v)	Choose Files s13207.v
Module Name	s13207
Netlist Type	Sequential ▾
Gate-level Library	LEDA ▾
Trigger Threshold	.2
No. Trigger Node	4
No. Rare Trigger Nodes	4
No. Trojans	1
Trojan Type	Template ▾
Template File (.v)	Choose Files troj_template.v
Clock	CK
Reset	Activation
SPF File (.spf)	Choose Files s13207.spf
	Submit

Fig. 3.8 Web interface for our DeTrust tool

Sequential The sequential Trojan consists of sample structure with q trigger conditions connected using an *and* gate. This trigger output feeds into a template 3-bit counter which executes the payload after the trigger condition activates 2^3 times. The Trojan-inserted design on the right of Fig. 3.10 is generated after the user inputs $\theta = 0.2$, $q = 3$, $r = 3$, $N = 1$, $type = $ seq, $effect = $ func, and $acti_mech = $ trigger.

Template The functional template Trojan option allows for users to insert their own Trojan structure with q customizable trigger conditions. Users must format the trigger inputs to their gate-level template so that they may be properly substituted and inserted. After running the tool with $\theta = 0.2$, $q = 4$ $N = 1$, $type = $ template, $effect = $ func, $activ_mech = $ trigger, and the template module above the Trojan is

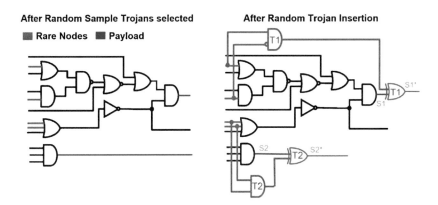

Fig. 3.9 Design after combinational Trojan insertion

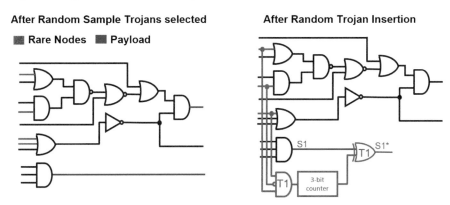

Fig. 3.10 Design after sequential Trojan insertion

inserted into the design shown in Fig. 3.11. The code snippets for the corresponding template before and after insertion are shown in Figs. 3.6 and 3.7.

For leakage Trojans, this tool allows for always-on or internally triggered activation mechanisms. Users must provide a template leakage circuit along with the critical information that is to be leaked. If critical signals are not specified, the tool will randomly select existing internal signals to leak. In case of always-on Trojans, all tool configurations regarding Trojan triggers are disregarded. Figure 3.12 shows a design with an always-on MOLES [14] template leaking three random internal signals with $N = 1$, $type =$ template, $effect =$ leakage, $activ\_mech =$ always_on.

Multiple Trojan Insertions   Users can also insert multiple Trojans in a design by specifying the configuration parameter $N > 1$ in Table 3.1. Inserting multiple Trojans can increase the threat level in certain scenarios. As a result, detecting one Trojan does not eliminate the possibility of another. For example, in the event that

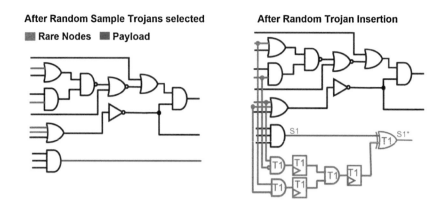

Fig. 3.11 Design after template Trojan insertion

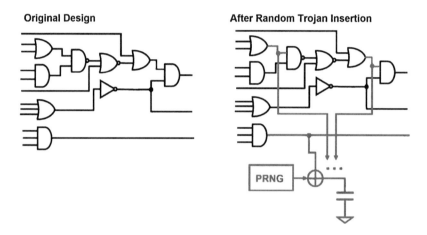

Fig. 3.12 Design after always-on leakage Trojan insertion

a functional Trojan is detected and removed from a third-party IP, another more stealthy leakage Trojan instance can remain. The existence of multiple Trojans in an IP can be more difficult and computationally expensive, especially when scaled to the System-on-Chip (SoC) level due to the increasing challenges in SoC verification [6].

Additional Configurations In addition to the configurations mentioned above, the current version of the tool also provides support for scan-chain insertions, clock definitions, and footprint optimizations. For scan chains, users must provide an SPF describing the scan-structure of the original design. Multiple clocks can be specified by providing the clock's signal name along with the activation level. For sequential or combinational Trojan types, if specified, the tool will seek to construct a Trojan with minimal switching activity and area from the provided specifications and original netlist through trigger and structure selection.

3.4.7 *Experimental Results and Analysis*

To demonstrate the effectiveness of the framework, we have inserted Trojans into ISCAS-85 and ISCAS-89 benchmarks and evaluate using a state-of-the-art Trojan detection methods: COTD [22]. A machine with Intel Core i5-3470 CPU @ 3.20GHz and 8 GB of RAM is used for testing. This tool supports flattened gate-level designs written in Verilog. The current implementation supports two standard cell libraries (LEDA and SAED). COTD is an unsupervised machine learning approach for Trojan detection. Sandia Controllability/Observability Analysis Program (SCOAP) controllability and observability values are extracted from nets in a gate-level netlist using Synopsys TetraMAX. The feature set includes a two-dimensional vector with combinational controllability and combinational observability (CC, CO) values which are clustered with k-means clustering using $k = 3$ in a simple Python script. During evaluation, we used sequential ISCAS benchmarks with combinational or sequential Trojans. For each benchmark, we insert Trojan instances with trigger nodes $(r/q) = 5/6$, 6/6, and 7/7. A low rare node threshold of $\theta = 0.0001$ was chosen for most benchmarks. θ is adjusted up to 0.05 when the tool is unable to generate feasible Trojans at $\theta = 0.0001$.

The results from evaluating COTD with Trojan-inserted designs from the tool are described in Table 3.2. We assume full-scan implementation for the original design and non-scan for Trojan insertions. The benchmarks are listed in the first column. The naming convention is B-Tq, where B is the original benchmark, T is the Trojan type (c, s, t for comb., seq., and template, respectively), and q is the number of trigger nodes. For example, s13207-c2 is the s13207 benchmark infected with a 2-node trigger combinational Trojan. If the number of rare nodes $r < q$, then the naming is B-Tr_q. The next three columns describe the tool configurations in terms of number trigger nodes, rare threshold value θ, and the Trojan type and number of instances. Columns 5 and 6 show the number of genuine signals and false negatives (FN) and the number of Trojan signals and false positives (FP). The last four columns provide information on the clustering itself by describing the cluster centroid for Trojan Cluster 1 (high CC values) and Trojan Cluster 2 (high CO values) and the minimum and maximum magnitude of the (CC, CO) pairs for genuine and Trojan signals. From Table 3.2, we can make the following observations:

1. In the presence of combinational Trojans, the COTD technique has a high false positive rate.
2. In the presence of sequential Trojans, the COTD technique has a low false positive rate.

While COTD does not produce any false negative signals, using Trojans generated from the tool caused COTD to generate false positives in all benchmarks. From the second observation, the false positives can be attributed to the payload gate's effect on the remaining circuit. The original design |CC, CO| for s13207, s15850, and s35932 are |1.414, 63.016|, |1.414, 139.717|, and |1.414, 12.018|, respectively. In the Trojan-inserted designs, the payload propagates its high CC and CO values to

Table 3.2 Compare COTD with random Trojan insertions

Benchmarks	Trig. nodes (r/q)	θ	Trojan (type, no.)	No. genuine signals (FN)	No. Trojan signals (FP)	Troj. Clstr. 1 Cntr <CC,CO>	Troj. Clstr. 2 Cntr <CC,CO>	Genuine Sig ICC,COI (min,max)	Troj. Sig ICC,COI (min,max)
s13207-c5_6	5/6	0.0001	comb (1)	2195(0)	544(538)	<7.158, 20.612>	<8.455, 44.743>	(1.414, 73.355)	(56.930, 70.093)
s13207-c6	6/6	0.0001	comb (1)	2197(0)	545(536)	<6.942, 20.665>	<8.888, 45.178>	(1.414, 63.016)	(62.777, 84.077)
s13207-c7	7/7	0.0001	comb (1)	2194(0)	547(539)	<7.461, 20.778>	<8.715, 46.264>	(1.414, 114.022)	(72.478, 112.060)
s13207-s5_6	5/6	0.0001	seq (1)	2730(0)	20(3)	<359.210, 191.25>	<23.810, 254>	(1.414, 254.387)	(254, 439.941)
s13207-s6	6/6	0.0001	seq (1)	2730(0)	19(3)	<347.563, 170.333>	<33.045, 253>	(1.414, 254.393)	(254, 439.941)
s13207-s7	7/7	0.0001	seq (1)	2730(0)	19(3)	<359.210, 191.25>	<33.752, 254>	(1.414, 254.387)	(254.330, 439.941)
s15850-c5_6	5/6	0.0001	comb (1)	2700(0)	731(722)	<16.031, 5.258>	<7.228, 19.615>	(1.414, 139.720)	(50.334, 64.101)
s15850-c6	6/6	0.0001	comb (1)	2912(0)	518(510)	<96.158, 22>	<8.604, 18.869>	(1.414, 139.718)	(82.644, 105.062)
s15850-c7	7/7	0.0001	comb (1)	2915(0)	515(507)	<111.679, 13>	<8.747, 19.033>	(1.414, 139.718)	(92.596, 121.103)
s15850-s5_6	5/6	0.0001	seq (1)	3419(0)	21(3)	<347.522, 170.333>	<24.476, 254>	(1.414, 254.026)	(254.098, 439.941)
s15850-s6	6/6	0.0001	seq (1)	3418(0)	21(4)	<359.210, 191.25>	<25.451, 254>	(1.414, 254.019)	(254.161, 439.941)
s15850-s7	7/7	0.0001	seq (1)	3419(0)	20(3)	<359.210, 191.25>	<23.961, 254>	(1.414, 254.019)	(254.051, 439.941)

s35932-c5_6	5/6	0.05	comb (1)	5185(0)	3173(3163)	<5.831, 4.777>	<2.371, 9.585>	(1.414, 50.020)	(31.969, 43.446)
s35932-c6	6/6	0.05	comb (1)	5185(0)	3173(3163)	<5.846, 4.782>	<2.384, 9.615>	(1.414, 59.017)	(36.064, 52.278)
s35932-c7	7/7	0.05	comb (1)	5185(0)	3173(3162)	<5.848, 4.783>	<2.386, 9.704>	(1.414, 61.016)	(37.216, 54.268)
s35932-s5_6	5/6	0.05	seq (1)	8341(0)	25(7)	<359.210, 191.5>	<10.911, 254>	(1.414, 254.098)	(254.020, 439.941)
s35932-s6	6/6	0.05	seq (1)	8341(0)	26(8)	<359.210, 191.5>	<11.711, 254>	(1.414, 254.128)	(254.098, 439.941)
s35932-s7	7/7	0.05	seq (1)	8341(0)	26(7)	<359.210, 191.5>	<13.410, 254>	(1.414, 254.128)	(254.098, 439.941)

FN is False Negative, FP is False Positive

its fan-in and fan-out. Moreover, the non-scan structure will automatically produce low controllability and observability values in TetraMAX from the complexity of sequential structures in ATPG tools [17]. From this, we can say COTD will most likely result in higher false positives in the presence of partial-scan designs—a common practice in industry to reduce area overhead and testing time compared to full-scan implementations. Additionally, the false positive rate may be affected in the cases where not all trigger nodes are rare which is the case for Trojans with r/q as 5/6.

Table 3.3 presents the results of applying MERO with $q = 2$ and $q = 4$ on Trojan designs from the tool. The first four columns are the same as in Table 3.2

Table 3.3 Compare MERO with random Trojan insertions

Benchmark	Trig. nodes (r/q)	θ	Trojan type (type, no.)	MERO				Rare nodes (rare/total)
				q = 2		q = 4		
				Trig Cov (%)	Troj Cov (%)	Trig Cov (%)	Troj Cov (%)	
c2670	–	0.05	–	100	95.724	98.717	89.781	30/1010
c2670-c1	1/1	0.05	comb (1)	100	96.881	98.764	93.975	30//1011
c2670-c2	2/2	0.05	comb (1)	100	96.850	98.606	93.770	32/1012
c2670-c3	3/3	0.05	comb (1)	100	95.106	98.440	90.036	33/1014
c2670-c4	4/4	0.05	comb (1)	100	95.138	98.163	89.367	34/1014
c3540	–	0.05	–	98.670	77.615	81.768	53.039	170/1184
c3540-c1	1/1	0.05	comb (1)	98.624	77.152	82.603	54.110	170/1185
c3540-c2	2/2	0.05	comb (1)	98.693	78.402	83.548	55.365	171/1186
c3540-c3	3/3	0.05	comb (1)	98.547	78.003	84.012	57.315	173/1188
c3540-c4	4/4	0.05	comb (1)	99.022	78.081	84.478	58.060	172/1187
c5315	–	0.05	–	99.717	87.746	88.107	57.347	99/2485
c5315-c1	1/1	0.05	comb (1)	99.769	88.101	88.126	56.112	99/2486
c5315-c2	2/2	0.05	comb (1)	99.728	88.375	88.156	57.522	100/2487
c5315-c3	3/3	0.05	comb (1)	99.738	88.164	88.251	58.547	100/2487
c5315-c4	4/4	0.05	comb (1)	99.727	87.471	87.595	56.288	100/2487
c6288	–	0.1	–	100	99.771	99.548	99.095	47/2448
c6288-c1	1/1	0.1	comb (1)	100	99.770	99.398	98.947	47/2449
c6288-c2	2/2	0.1	comb (1)	100	99.744	99.554	99.088	48/2450
c6288-c3	3/3	0.1	comb (1)	100	99.741	99.035	98.232	48/2450
c6288-c4	4/4	0.1	comb (1)	100	99.773	99.541	98.928	48/2450
s13207	–	0.2	–	17.534	3.288	0	0	1355/2504
s13207-s1	1/1	0.2	seq (1)	19.710	3.043	0	0	1357/2514
s13207-s2	2/2	0.2	seq (1)	17.797	2.404	0	0	1358/2515
s13207-s3	3/3	0.2	seq (1)	17.226	2.92	6.667	0	1363/2517
s13207-s4	4/4	0.2	seq (1)	16.208	2.315	5.556	0	1362/2517

– is not applicable

which report the benchmark and the tool configurations. Columns 5–8 show the trigger and Trojan coverage. The last column presents the number of rare nodes reported from MERO over the total nodes. The low trigger and Trojan coverage for the sequential benchmark s13207 can be attributed to the non-scan implementation being used. In benchmarks c3540, c5315, and s13207 the original benchmarks generally have lower trigger and Trojan coverage than the Trojan-inserted designs. For the remaining benchmarks, the original benchmarks have higher trigger and Trojan coverage than the Trojan-inserted designs. However, only the cases in which the number of trigger nodes for MERO matched the number of trigger nodes in the Trojan, MERO is likely to detect the Trojan. Therefore, MERO would only benefit from detecting small Trojan insertions as increasing q increases runtime.

3.5 Summary

IP metrics and benchmarks are an important tool in the evaluation of novel techniques developed by researchers. In hardware security and trust, existing static trust benchmarks for Trojan detection provide a good foundation but are limited in Trojan variety and robustness. They also do not evolve with new attack modalities being discovered. We have presented a comprehensive automatic Trojan insertion framework with associated algorithms that provide users the ability to generate dynamic benchmarks with random Trojan insertions. Users can control several parameters regarding Trojan trigger, structure, payload type, and rarity. Additionally, future Trojan models are supported by allowing template Trojans. This tool can be used to evaluate Trojan detection methods from a red team vs blue team perspective demonstrated using a popular Trojan detection method.

References

1. S. Amir et al., Development and evaluation of hardware obfuscation benchmarks. J. Hardw. Syst. Secur. **2**, 142–161 (2018)
2. S. Bhunia et al., Hardware Trojan attacks: threat analysis and countermeasures, in *IEEE Special Issue on Trustworthy Hardware* (2014)
3. M. Bilzor, T. Huffmire, C. Irvine, T. Levin, Evaluating security requirements in a general purpose processor by combining assertion checkers with code coverage, in *Hardware-Oriented Security and Trust (HOST)* (2012)
4. R. Chakraborty et al., MERO: a statistical approach for hardware Trojan detection, in *International Workshop on Cryptographic Hardware and Embedded Systems (CHES'09)* (2009)
5. S. Charles, Y. Lyu, P. Mishra, Real-time detection and localization of DoS attacks in NoC based SoCs, in *Design Automation and Test in Europe (DATE), Florence, Italy, March 25–29* (2019)
6. M. Chen et al., *System-Level Validation: High-Level Modeling and Directed Test Generation Techniques* (Springer, New York, 2012)

7. J. Cruz, Y. Huang, P. Mishra, S. Bhunia, An automated configurable Trojan insertion framework for dynamic trust benchmarks, in *Design Automation and Test in Europe (DATE), Dresden, Germany, March 19–23* (2018), pp. 1598–1603
8. J. Cruz, P. Mishra, S. Bhunia, The metric matters: how to measure trust, in *Design Automation Conference (DAC), Las Vegas, June 2–6* (2019)
9. F. Farahmandi, Y. Huang, P. Mishra, Trojan localization using symbolic algebra, in *Asia and South Pacific Design Automation Conference (ASP-DAC)* (2017), pp. 591–597
10. X. Guo et al., Pre-silicon security verification and validation: a formal perspective, in *DAC* (2015)
11. M. Hicks, M. Finnicum, S.T. King, M.M.K. Martin, J.M. Smith, Overcoming an untrusted computing base: detecting and removing malicious hardware automatically, in *Proceedings of IEEE Symposium on Security and Privacy* (2010), pp. 159–172
12. Y. Huang, S. Bhunia, P. Mishra, MERS: statistical test generation for side-channel analysis based Trojan detection, in *ACM Conference on Computer and Communications Security (CCS)* (2016)
13. Y. Huang, S. Bhunia, P. Mishra, Scalable test generation for Trojan detection using side channel analysis. IEEE Trans. Inf. Forensics Secur. **13**(11), 2746–2760 (2018)
14. L. Lang et al., MOLES: malicious off-chip leakage enabled by side-channels, in *ICCAD* (2009)
15. E. Love, Y. Jin, Y. Makris, Proof-carrying hardware intellectual property: a pathway to trusted module acquisition. IEEE Trans. Inf. Forensics Secur. **7**(1), 2540 (2012)
16. Y. Lyu, P. Mishra, Efficient test generation for Trojan detection using side channel analysis, in *Design Automation and Test in Europe (DATE), Florence, Italy, March 25–29* (2019)
17. T.E. Marchok et al., Complexity of sequential ATPG, in *DATE* (1995)
18. P. Mishra et al., *Hardware IP Security and Trust* (Springer, Cham, 2016)
19. A. Nahiyan, M. Tehranipoor, Code coverage analysis for IP trust verification, in *Hardware IP Security and Trust*, ed. by P. Mishra, S. Bhunia, M. Tehranipoor (Springer, Cham, 2017). ISBN 978-3-319-49025-0
20. S. Narasimhan, R.S. Chakraborty, S. Chakraborty, Hardware IP protection during evaluation using embedded sequential Trojan. IEEE Des. Test Comput. **29**(3), 70–79 (2012)
21. J. Rajendran, O. Sinanoglu, R. Karri, Building trustworthy systems using untrusted components: a high-level synthesis approach. IEEE Trans. Very Large Scale Integr. VLSI Syst. **24**(9), 2946–2959 (2016)
22. H. Salmani, COTD: reference-free hardware Trojan detection and recovery based on controllability and observability in gate-level netlist. IEEE Trans. Inf. Forensics Secur. **12**, 338–350 (2016)
23. H. Salmani, M. Tehranipoor, Analyzing circuit vulnerability to hardware Trojan insertion at the behavioral level, in *IEEE International Symposium on Defect and Fault Tolerance in VLSI and Nanotechnology Systems (DFT)* (2013), pp. 190–195
24. H. Salmani, M. Tehranipoor, R. Karri, On design vulnerability analysis and trust benchmark development, in *IEEE International Conference on Computer Design (ICCD)* (2013)
25. B. Shakya, T. He, H. Salmani, D. Forte, S. Bhunia, M. Tehranipoor, Benchmarking of hardware Trojans and maliciously affected circuits. J. Hardw. Syst. Secur. **1**, 85–102 (2017)
26. C. Sturton, M. Hicks, D. Wagner, S. King, Defeating UCI: building stealthy and malicious hardware, in *2011 IEEE Symposium on Security and Privacy (SP)* (2011), pp. 64–77
27. Synopsys, TetraMAX ATPG User Guide, Version H-2013.03-SP4 (2013)
28. M. Tehranipoor, F. Koushanfar, A survey of hardware Trojan taxonomy and detection. IEEE Des. Test Comput. **27**(1), 10–25 (2010)
29. M. Tehranipoor, H. Salmani, X. Zhang, *Integrated Circuit Authentication: Hardware Trojans and Counterfeit Detection* (Springer, New York, 2013)
30. A. Waksman, M. Suozzo, S. Sethumadhavan, FANCI: identification of stealthy malicious logic using Boolean functional analysis, in *Proceedings of the ACM Conference on Computer & Communications Security* (2013), pp. 697–708

31. Y. Xie, A. Srivastava, Mitigating SAT attack on logic locking, in *Proceedings of 18th International Conference on CHES 2016, Santa Barbara, CA, USA* (Springer, Berlin, 2016), pp. 127–146
32. M. Yasin et al., On improving the security of logic locking. IEEE Trans. Comput. Aided Des. Integr. Circuits Syst. **35**(9), 1411–1424 (2016)
33. M. Yasin, B. Mazumdar, J.J.V. Rajendran, O. Sinanoglu, SARLock: SAT attack resistant logic locking, in *IEEE International Symposium on HOST 2016, McLean, VA, USA* (2016), pp. 236–241

Chapter 4
Anomaly Detection Using Symbolic Algebra

4.1 Introduction

The urge of high speed and high precision computations increases use of arithmetic circuits in real-time applications such as multimedia and cryptography operations. Optimized and custom arithmetic architectures are required to meet these high speed and precision constraints. There is a critical need for efficient arithmetic circuit security verification techniques due to Trojan proneness of non-standard arithmetic circuit implementations. Hence, the automated security verification of arithmetic circuits is absolutely necessary for efficient design validation.

A major problem with design validation is that we do not know whether an anomaly exists, and how to quickly detect and fix it. We can always keep on generating random tests, in the hope of activating the malicious functionality; however, random test generation is neither scalable nor efficient when designs are large and complex. Existing directed test generation techniques [1, 2, 8–12, 18, 24, 28, 29, 31–33] are promising only when the list of anomalies is available. In this chapter, we present a directed test generation technique that is guaranteed to activate unknown bugs (if any). The generated tests would also help for faster bug localization.

From the security point of view, it is important to prove that the design implementation is equivalent to its specification—nothing more, nothing less. Any deviation from the specification may endanger the correct functionality, trustworthiness, and the security of the design. Therefore, any errors in hardware designs threaten the integrity and security of the overall design. Notably, gate-replacement errors in the gate-level netlist can change the correct functionality of design and insert anomaly in its implementation. Moreover, gate-replacement error may pose security threats since it can act as a bit-flip (in comparison with the golden behavior) and cause unauthorized transitions to protected states of the design, wrong results, and denial of service. The situation gets worse when we consider such anomalies in arithmetic circuits that provide the result of encryption operations. Any error in the outcome

© Springer Nature Switzerland AG 2020
F. Farahmandi et al., *System-on-Chip Security*,
https://doi.org/10.1007/978-3-030-30596-3_4

of such secure operation may threaten the security of the overall design. Gate-replacement anomalies are small malicious modifications and have negligible effect on physical characteristics (area, power, and energy) of the design. Therefore, they cannot be detected during design review. Moreover, they cannot be easily activated using random and constraint-random validation approaches. Considering the huge complexity of SoCs' gate-level netlist, gate-replacement errors are likely due to the presence of rouge designers, EDA tool vulnerabilities, and optimization procedures.

Equivalence checking techniques seem promising to identify gate-replacement errors. However, existing equivalence checking methods can lead to state space explosion when complex and large IPs such as arithmetic circuits are involved. Existing arithmetic circuits verification approaches have focused on checking the equivalence between the specification of a circuit and its implementation. They use an algebraic model of the implementation [13, 26] using a set of polynomials F. The specification of an arithmetic circuit can be modeled as a polynomial f_{spec} using decimal representation of primary inputs and primary outputs. The verification problem is formulated as mathematical manipulation of f_{spec} over polynomials in F. If the implementation is equivalent to the specification, the result of equivalence checking is a zero polynomial; otherwise, it produces a polynomial containing primary inputs as variables. We call this polynomial *remainder*. Any assignment to remainder's variables that makes the remainder to have a non-zero decimal value generates one counter-example. Based on the location of a bug, remainder generation may be expensive. However, the remainder generation is one time effort and multiple counterexamples (directed tests) can be generated from one remainder.

Figure 4.1 shows different scenarios of a Trojan-inserted implementation. Figure 4.1a illustrates the case when only one Trojan exists in the implementation. Figure 4.1b shows the presence of two Trojans which do not share input cones (Trojan with independent triggers which we call them independent Trojans). We describe how to fix one or more independent Trojans in Sects. 4.3 and 4.4.1, respectively. We refer to "activation independence" in the context of activating independent Trojans as shown in Fig. 4.1b. In other words, independent Trojans do not have any overlapping input cones. On the other hand, their effect can be seen in different primary outputs. The effect of different Trojans (dependent

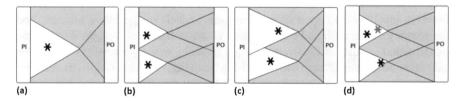

Fig. 4.1 Illustrative malicious scenarios for a given design with Trojan-specific input and output cones. Each star represents one Trojan. Here, PI and PO refer to the primary inputs and primary outputs, respectively. (**a**) Single bug. (**b**) Two independent bugs. (**c**) Two dependent bugs. (**d**) Hybrid of dependent and independent scenarios

Fig. 4.2 Overview of
different steps of the
presented security verification
framework. Independent
Trojans are located and
corrected using the first loop
(shown with dashed line) as
described in Sect. 4.4.1.
Detection/correction of
dependent Trojans is
discussed in Sect. 4.4.2

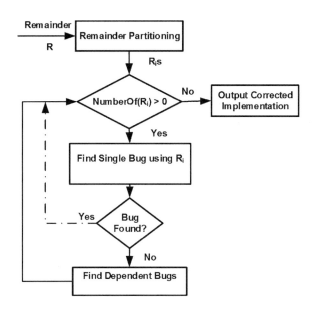

and independent Trojans) may appear in overlapping or nonoverlapping cones as
shown in Fig. 4.1. We present algorithms to locate and correct multiple independent
Trojans. In many cases, Trojans may share input cones (Trojans that have dependent
Trojans) as shown in Fig. 4.1c. In this chapter, we also present an algorithm to
automatically fix multiple dependent Trojans in Sect. 4.4.2. Generally, a Trojan-
inserted implementation can contain any combination of independent and dependent
malicious functionality as shown in Fig. 4.1d.

Figure 4.2 shows different steps of the presented equivalence checking approach
to locate and correct multiple Trojans for various scenarios depicted in Fig. 4.1. The
existence of a non-zero remainder as a result of applying the functional verification
between specification and implementation of an arithmetic circuit is a sign of an
untrustworthy implementation. However, there is no information about the number
of existing Trojans in the implementation. There can be a single Trojan or multiple
independent/dependent Trojans in the design. In Sect. 4.3, we present a single
Trojan correction algorithm. The main question is that how to know the number
of remaining Trojans in the design and which algorithm should be used to fix them.
In order to determine whether there is more than one Trojan in the implementation,
we try to partition the remainder R into sub-remainders R_i first. If the remainder can
be partitioned successfully into n sub-remainders, we can conclude that there are at
least n independent Trojans in the implementation as we discussed in Sect. 4.4.1.
Algorithms in Sect. 4.3 are used over each sub-remainder R_i to fix each Trojan.
However, if a single Trojan cannot be found for remainder R_i, there are multiple
dependent Trojans which construct the sub-remainder R_i. Therefore, we try to find
a single Trojan corresponding to remainder R_i first. If we can find such a Trojan, the
Trojan will be fixed. Otherwise, we try the proposed algorithm of Sect. 4.4.2 to find

dependent Trojans responsible for sub-remainder R_i. The procedure is repeated for all of the sub-remainders.

The remainder of this chapter is organized as follows: Section 4.2 reviews symbolic algebra and how it can be used for verification purposes. Section 4.3 discusses the framework for directed test generation and Trojan localization/detection approach. Section 4.4 describes the security verification approach to detect and correct multiple Trojans. Section 4.5 describes challenges in remainder generation and presents an approach to address them. Section 4.6 presents the experimental results. Finally, Sect. 4.7 concludes this chapter.

4.2   Fundamental of Verification Using Symbolic Algebra

In this section, we briefly describe Gröbner basis theory [14]. Next, we present the application of Gröbner basis theory for verification of hardware designs.

4.2.1   Gröbner Basis Theory

Let $M = x_1^{\alpha_1} x_2^{\alpha_2} \dots x_n^{\alpha_n}$ be a monomial and $f = C_1 M_1 + C_2 M_2 + \dots + C_t M_t$ be a polynomial with $\{c_1, c_2, \dots, c_t\}$ as coefficients and $M_1 > M_2 > \dots > M_t$. Monomial $lm(f) = M_1$ is called leading monomial and $lt(f) = C_1 M_1$ is called leading term of polynomial f. Let $\mathbb{K}$ be a computable field and $\mathbb{K}[x_1, x_2, \dots, x_n]$ be a polynomial ring in n variables. Then $< f_1, f_2, \dots, f_s > = \{ \sum_{i=1}^{n} h_i f_i :$ $h_1, h_2, \dots, h_s \in \mathbb{K}[x_1, x_2, \dots, x_n]\}$ is an ideal I. The set $\{f_1, f_2, \dots, f_s\}$ is called generator or basis of ideal I. If $V(I)$ shows the affine variety (set of all solutions of $f_1 = F_2 = \dots = f_s = 0$) of ideal I, $I(V) = \{f_i \in \mathbb{K}[x_1, x_2, \dots, x_n] :$ $\forall v \in V(I), f_i(v) = 0\}$. Polynomial f_i is a member of $I(V)$ if it vanishes on $V(I)$. Gröbner basis is one of the generators of every ideal I (when I is other than zero) that has a specific characteristic to answer membership problem of an arbitrary polynomial f in ideal I. The set $G = \{g_1, g_2, \dots, g_t\}$ is called Gröbner basis of ideal I, if $\forall f_i \in I, \exists g_j \in G : lm(g_j)|lm(f_i)$.

The Gröbner basis solves the membership testing problem of an ideal using sequential divisions or reduction. The reduction operation can be formulated as follows. Polynomial f_i can be reducible by polynomial g_j if $lt(f_i) = C_1 M_1$ (which is non-zero) is divisible by $lt(g_i)$ and r is the remainder ($r = f_i - \frac{lt(f_i)}{lt(g_j)} \cdot g_j$). It can be denoted by $f_i \xrightarrow{g_j} r$. Similarly, f_i can be reducible with respect to set G and it can be represented by $f_i \xrightarrow{G}_+ r$.

The set G is Gröbner basis ideal I, if $\forall f \in I, f_i \xrightarrow{G}_+ 0$. Gröbner basis can be computed using Buchberger's algorithm [5]. Buchberger's algorithm is shown in

---

Algorithm 2: Buchberger's algorithm [5]

Input: ideal $I = < f_1, f_2, \ldots, f_s > \neq \{0\}$, initial basis $F = \{f_1, f_2, \ldots, f_s\}$
Output: Gröbner Basis $G = \{g_1, g_2, \ldots, g_t\}$ for ideal I
$G = F$
$V = G \times G$
while $V \neq 0$ **do**
    for *each pair* $(f, g) \in V$ **do**
        $V = V - (f, g)$
        $Spoly(f, g) \rightarrow_G r$
        if $r \neq 0$ **then**
            $G = G \cup r$
            $V = V \cup (G \times r)$
        end
    end
end
return *set* G

Algorithm 2. It makes use of a polynomial reduction technique named S-polynomial as defined below.

Definition 4.1 (S-polynomial) Assume $f, g \in \mathbb{K}1, x_2, \ldots, x_n]$ are non-zero polynomials. The S-polynomial of f and g (a linear manipulation of f and g) is defined as: $Spoly(f, g) = \frac{LCM(LM(f), LM(g))}{LT(f)*f} - \frac{LCM(LM(f), LM(g))}{LT(g)*g}$, where $LCM(a, b)$ is a notation for the least common multiple of a and b.

Example 4.1 Let $f = 6*x_1^4*x_2^5 + 24*x_1^2 - x_2$ and $g = 2*x_1^2*x_2^7 + 4*x_2^3 + 2*x_3$ and we have $x_1 > x_2 > x_3$. The S-polynomial of f and g is defined below:

$$LM(f) = x_1^4 * x_2^5$$

$$LM(g) = x_1^2 * x_2^7$$

$$LCM(x_1^4 * x_2^5, x_1^2 * x_2^7) = x_1^4 * x_2^7$$

$Spoly(f, g) = \frac{x_1^4*x_2^7}{6*x_1^4*x_2^5)} * f - \frac{x_1^4*x_2^7}{2*x_1^2*x_2^7} * g = 4x_1^2 * x_2^2 - \frac{1}{6} * x_2^3 - 2 * x_1^2 * x_2^3 - x_1^2 * x_3$

It is obvious that S-polynomial computation cancels leading terms of the polynomials. As shown in Algorithm 2, Buchberger's algorithm first calculates all S-polynomials (lines 4–6) and then adds non-zero S-polynomials to the basis G (line 8). This process repeats until all of the computed S-polynomials become zero with respect to G. It is obvious that Gröebner basis can be extremely large so its computation may take a long time and it may need large storage memory as well. The time and space complexity of this algorithm are exponential in terms of the sum of the total degree of polynomials in F, plus the sum of the lengths of the

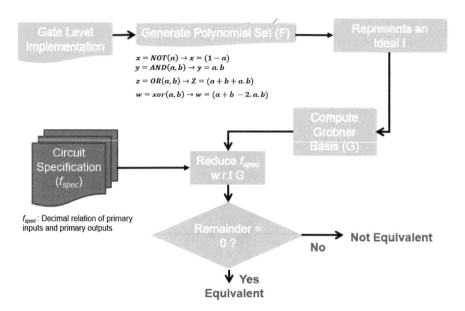

$$x = NOT(a) \rightarrow x = (1 - a)$$
$$y = AND(a, b) \rightarrow y = a.b$$
$$z = OR(a, b) \rightarrow Z = (a + b + a.b)$$
$$w = xor(a, b) \rightarrow w = (a + b - 2.a.b)$$

f_{spec}: Decimal relation of primary
inputs and primary outputs

Fig. 4.3 Equivalence checking using symbolic algebra

polynomials in F [5]. When the size of F increases, the verification process may be
very slow or in the worst-case may be infeasible.

Buchberger's algorithm is computationally intensive and it may affect the
performance drastically. It has been shown in [6] that if every pair (f_i, f_j) that
belongs to set $F = \{f_1, f_2, \ldots, f_s\}$ (generator of ideal I) has a relatively prime
leading monomials $(lm(f_i).lm(f_j) = LCM(lm(f_i).lm(f_j)))$ with respect to order
$>$, the set F is also Gröbner basis of ideal I.

Based on these observations, efficient equivalence checking between specification of an arithmetic circuit and its implementation can be performed as shown in
Fig. 4.3. The major computation steps in Fig. 4.3 are outlined below:

- Assuming a computational field $\mathbb{K}$ and a polynomial ring $\mathbb{K}[x_1, x_2, \ldots, x_n]$ (note
  that variables $\{x_1, x_2, \ldots, x_n\}$ are subset of signals in the gate-level implementation), a polynomial $f_{spec} \in \mathbb{K}[x_1, x_2, \ldots, x_n]$ representing specification of the
  arithmetic circuit can be derived.
- Map the implementation of arithmetic circuit to a set of polynomials that belongs
  to $\mathbb{K}[x_1, x_2, \ldots, x_n]$. The set F generates an ideal I. Note that according to the
  field $\mathbb{K}$, some vanishing polynomials that constructs ideal I_0 may be considered
  as well.
- Derive an order $>$ in a way that leading monomials of every pair (f_i, f_j) are
  relatively prime. Thus, the generator set F is also Gröbner basis $G = F$. As the
  combinational arithmetic circuits are acyclic, the topological order of the signals
  in the gate-level implementation can be used.

- The final step is reduction of f_{spec} with respect to Gröbner basis G and order $>$.

  In other words, the verification problem is formulated as $f_{spec} \xrightarrow{G}_+ r$. The gate-level circuit C has correctly implemented specification f_{spec} if the remainder r is equal to 0. The non-zero remainder implies a gate-replacement Trojan in the implementation.

Galois field arithmetic computation can be seen in Barrett reduction [22], Mastrovito multiplication and Montgomery reduction [23] which are critical part of cryptosystems. In order to apply the method of Fig. 4.3 for verification of Galois field arithmetic circuits, strong Nullstellensatz over Galois fields can be used. Galois field is not an algebraically closed field, so its closure should be used. Strong Nullstellensatz helps to construct a radical ideal in a way such that $I(V_{\mathbb{F}_{2^k}}) = I + I_0$. Ideal I_0 is constructed by using vanishing polynomials $x_i^{2^k} - x_i$ by considering the fact that $\forall x_i^{2^k} \in \mathbb{F}_{2^k} : x_i^{2^k} - x_i = 0$. As a result, the Gröbner basis theory can be applied on Galois field arithmetic circuits. The method in [26] has extracted circuit polynomials by converting each gate to a polynomial, and SINGULAR [21] has been used to do the $f_{spec} \xrightarrow{G}_+ r$ computations. Using this method, the verification of Galois field arithmetic circuits like Mastrovito multipliers with up to 163 bits can be done in a few hours. Some extensions of this method have been proposed in [27]. The cost of $f_{spec} \xrightarrow{G}_+ r$ computation has been improved by mapping the computation on a matrix representing the verification problem, and the computation is performed using Gaussian elimination.

The Gröbner basis theory has been used to verify arithmetic circuits over ring $\mathbb{Z}[x_1, x_2, \ldots, x_n]/2^N$ in [19]. Instead of mapping each gate to a polynomial, the repetitive components of the circuit are extracted and the whole component is represented using one polynomial (since arithmetic circuit over ring $\mathbb{Z}[x_1, x_2, \ldots, x_n]/2^N$ contains carry chain, the number of polynomials can be very large). Therefore, the number of circuit polynomials is decreased. In order to expedite the $f_{spec} \xrightarrow{G}_+ r$ computation, the polynomials are represented by Horner expansion diagrams [19]. The reduction computation is implemented by sequential division. The verification of arithmetic circuit over ring $\mathbb{Z}[x_1, x_2, \ldots, x_n]/2^N$ up to 128 bit can be efficiently performed using this method. An extension of this method has been presented in [15] that is able to significantly reduce the number of polynomials by finding fanout-free regions and representing the whole region by one single polynomial. Similar to [27], the reduction of specification polynomial with respect to Gröbner basis polynomials is performed by Gaussian elimination resulting in verification time of few minutes. In all of these methods, when the remainder r is non-zero, it shows that the specification is not exactly equivalent with the gate-level implementation. Thus, the non-zero remainder can be analyzed to identify the hidden malfunctions or Trojans in the system. In this section, the use of one of these approaches for equivalence checking of integer arithmetic circuits over $\mathbb{Z}_{2^n}$ is explained. Although the details are different for Galois Field arithmetic circuits, the major steps are similar.

4.2.2 Verification of Arithmetic Circuits

Most of the traditional verification and debugging tools of arithmetic circuits are based on techniques such as simulation, binary decision diagrams (like BDDs,*BMD [4]), and SAT solvers [25, 30]. However, all of these approaches suffer from state space explosion while dealing with large and complex circuits especially arithmetic circuits. Furthermore, most of these approaches cannot provide concrete suggestions to remove Trojans. It is important to introduce efficient, scalable, and fully automated verification framework.

Computer symbolic algebra is employed for equivalence checking of arithmetic circuits to address the limitations of traditional approaches. The primary goal is to check equivalence between the specification polynomial f_{spec} and gate-level implementation C to find potential malicious functionality. The specification of arithmetic circuit and implementation is formulated as polynomials. Arithmetic circuits constitute a significant portion of datapath in signal processing, cryptography, multimedia applications, error root causing codes, etc. In most of them, arithmetic circuits have a custom structure and can be very large so the chances of potential malfunction are high. These Trojans may cause unwanted operations as well as security problems like leakage of secret key [3]. Thus, verification of arithmetic circuits is very important.

The arithmetic circuit equivalence checking problem formulation starts with converting the design specification to a polynomial f_{spec} which represents the word-level abstraction of arithmetic circuits functionality using primary inputs and primary outputs as variables. For example, the specification of an n-bit adder with primary inputs $A = \{a_0, a_1, \ldots, a_{n-1}\}$ and $B = \{b_0, b_1, \ldots, b_{n-1}\}$ and primary output $Z = \{z_0, z_1, \ldots z_n\}$ can be formulated as $Z = A + B$ or can be written as $(2^n.z_n + \ldots + 2.z_1 + z_0) - ((2^{n-1}.a_{n-1} + \ldots + 2.a_1 + a_0) + (2^{n-1}.b_{n-1} + \ldots + 2.b_1 + b_0)) = 0$, where $\{a_i, b_i, z_i\} \subset \{0, 1\}$.

The functionality of logic gates (such as AND, OR, XOR, NOT, and buffer) can be represented by polynomials such that the input and output signals of gates act as variables of the corresponding polynomial. Each variable x_i which appears in a circuit polynomial belongs to $\mathbb{Z}_2$, where $(x_i^2 = x_i)$. Equation 4.1 shows the corresponding polynomial of NOT, AND, OR, XOR gates. Note that any complex gate can be modeled as a combination of these gates and its polynomial can be computed by combining the equations shown in Eq. 4.1.

$$
\begin{aligned}
z_1 &= \text{NOT}(a) \rightarrow z_1 = 1 - a, \\
z_2 &= \text{AND}(a, b) \rightarrow z_2 = a.b, \\
z_3 &= \text{OR}(a, b) \rightarrow z_3 = a + b - a.b, \\
z_4 &= \text{XOR}(a, b) \rightarrow z_4 = a + b - 2.a.b
\end{aligned}
\tag{4.1}
$$

A gate-level netlist of a circuit can be modeled as a set of polynomials $\mathbb{F}$ by modeling each gate as a polynomial. Suppose that we want to make sure an arithmetic circuit implements correctly its specification. In other words, we want to verify that there are no functional errors in the arithmetic circuit. The equivalence checking starts with consecutively reducing the f_{spec} over implementation polynomials ($\mathbb{F}_{imp}$) until either zero remainder or a remainder that contains only primary input variables is reached. If the remainder is zero, it shows that the arithmetic circuit performs the exact specification. However, the non-zero remainder shows that the implementation is not trustworthy and there are some malfunctions.

Example 4.2 Suppose that we want to verify the functional correctness of a full-adder implementation shown in Fig. 4.4. The specification can be formulated as: $(2.C_{out} + S - (A + B + C_{in}))$ and each gate in the implementation can be modeled as a polynomial based on Eq. 4.1. The topological order of the circuit (since the circuit is acyclic) is chosen for reduction as $C_{out} > \{S, n_3\} > \{n_2, n_1\} > \{A, B, C_{in}\}$. The reduction starts from the most significant primary output and ends at primary inputs. Variables in the curvy brackets have the same order and they can be reduced in one iteration. Equation 4.2 shows the reduction process. It can be seen that the final result (remainder) is a non-zero polynomial and implementation is not trustworthy. It is easy to verify that the remainder would be zero if the NAND gate is replaced with an AND gate. ∎

$$step_0 : 2.C_{out} + S - A - B - C_{in}$$

$$step_1 : S - 2.n_3.n_2 + 2.n_3 + 2.n_2 - A - B - C_{in}$$

$$step_2 : 2.n_2.n_1.C_{in} - 4.n_1.C_{in} + n_1 - A - B + 2$$

$$step_3(remainder) : 8.A.B.C_{in} - 4.A.C_{in} - 4.B.C_{in} - 2.A.B + 2$$

(4.2)

Fig. 4.4 The tampered gate-level netlist of a full-adder. The NAND gate should be replaced by an AND gate to correct the Trojan

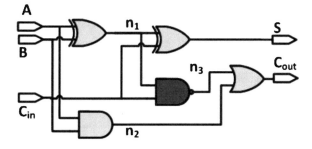

4.3   Automated Security Verification Using Remainders

The framework uses the remainder that is generated by equivalence checking. If the remainder is a non-zero polynomial, it means that the implementation is Trojan-inserted; however, the source of the Trojan is unknown.

Example 4.3 Consider a 2-bit multiplier with gate-level netlist shown in Fig. 4.5. Suppose that we deliberately insert a Trojan in the circuit shown in Fig. 4.5 by putting the XOR gate with inputs (A_0, B_0) instead of an AND gate. The specification of a 2-bit multiplier is shown by f_{spec}. The verification process starts from f_{spec} and replaces its terms one by one using information derived from the implementation polynomials as shown in Eq. 4.3. For instance, term $4.Z_2$ from f_{spec} is replaced with expression $(R + O - 2.R.O)$. The topological order $\{Z_3, Z_2\} > \{Z_1, R\} > \{Z_0, M, N, O\} > \{A_0, A_1, B_0, B_1\}$ is considered to perform term rewriting. The verification result is shown in Eq. 4.3. Clearly, the remainder is a non-zero polynomial and it reveals the fact that the implementation is not trustworthy.   ■

$$f_{spec} : 8.Z_3 + 4.Z_2 + 2.Z_1 + Z_0 - 4.A_1.B_1 - 2.A_1 B_0 - 2.A_0.B_1 - A_0.B_0$$

$$step_1 : 4.R + 4.O + 2.z_1 + Z_0 - 4.A_1.B_1 - 2.A_1 B_0 - 2.A_0.B_1 - A_0.B_0$$

$$step_2 : 4.O + 2.M + 2.N + Z_0 - 4.A_1.B_1 - 2.A_1 B_0 - 2.A_0.B_1 - A_0.B_0$$

$$step_3(remainder) : 1.A_0 + 1.B_0 - 3.A_0.B_0$$

$$(4.3)$$

This approach takes the remainder and the implementation as inputs and tries to find the source of Trojan in the implementation and correct it. The presented framework has three important steps. First, we use the remainder to generate directed tests to activate malicious scenarios. Next, we try to localize source of the Trojan by leveraging the generated tests. Finally, we use an automated correction technique to detect and correct the existing Trojan which resides in the suspicious area. We describe each of these steps in detail in the following sections.

Fig. 4.5 The Trojan-inserted gate-level netlist of a 2-bit multiplier

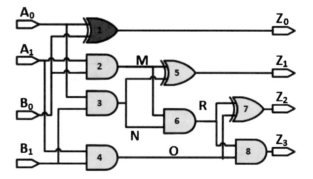

It has been shown that if and only if the remainder is zero, the implementation is Trojan-free [34]. Thus, when we have a non-zero polynomial as a remainder, any assignment to its variables that makes the numerical value of the remainder non-zero is a test to activate the Trojan(s). In the presented approach, we make use of the remainder to generate test cases to activate unknown Trojans. The test is guaranteed to activate the Trojan in the design. The remainder is a polynomial with Boolean/integer coefficients. It contains a subset of primary inputs as its variables. The approach takes the remainder and finds the possible assignments to its variables such that it makes the numerical value of the remainder non-zero. As shown in Example 4.3, the remainder may not contain all of the primary inputs. As a result, the approach may use a subset of the primary inputs (that appear in the remainder) to generate directed tests with "don't cares." Such assignments can be found using a SMT solver by defining Boolean variables and considering signed/unsigned integer values as the total value of the remainder polynomial ($i \neq 0 \in \mathbb{Z}$, $check(R = i)$). The problem of using SMT solver is that for each i, it finds at most one assignment of the remainder variables to produce value of i, if possible. We implemented an optimized algorithm to find all possible assignments that produce non-zero numerical values of the remainder. Algorithm 3 shows the details of the test generation method. The algorithm takes remainder (R) polynomial as well as primary inputs (PI) as inputs and generates a set of directed tests $\mathbb{T}$ to activate the Trojan. A remainder is constructed as a set of terms as $R = T_1 + T_2 + \ldots + T_n$, where each term T_j is a product of a coefficient C_j and a monomial M_j. The algorithm tries different sets of binary values to PIs (s_i)s, and computes the numerical value of R for assignment s_i. M_i is a product of binary variables. The value of M_j is either one or zero as it is a product of some binary variables (line 7). Therefore, the term value may be zero or equal to the term coefficient (C_j). To compute the numerical value of R for assignment s_i, the algorithm computes the sum of the values of all the terms in the remainder (lines 4–8). If the sum of all the terms is non-zero, the corresponding primary input assignments are added to the set of tests (lines 9–10). The test generation algorithm can be implemented in a parallel fashion to improve its performance (Table 4.1).

Example 4.4 Consider the tampered circuit shown in Fig. 4.5 and the remainder polynomial $R = A_0 + B_0 - 3.A_0.B_0$. The assignments that make R to have a non-zero numerical value ($R = 1$ or $R = -1$) are ($A_0 = 1$, $B_0 = 0$), ($A_0=0$, $B_0=1$), and ($A_0 = 1$, $B_0 = 1$). These are the scenarios that make difference between functionality of an AND gate and an XOR gate. Otherwise, the fault will be masked since when ($A_0 = 0$, $B_0 = 0$), AND and XOR produce the same output.   ∎

4.3.1   Trojan Localization

So far, we know that the implementation is untrustworthy and we have all the necessary tests to activate the malicious scenarios. The goal is to reduce the state space in order to localize the Trojan by using the tests generated in the previous

Table 4.1 Directed tests to
activate the Trojan shown in
Fig. 4.5

A_1	A_0	B_1	B_0
X	1	X	0
X	0	X	1
X	1	X	1

Algorithm 3: Directed test generation algorithm

Input: Remainder, R
Output: Directed Tests $\mathbb{T}$
for *different assignments s_i of PIs in R* **do**
  Sum = 0
  for *each term $T_j = C_j.M_j \in R$* **do**
    if $(M_j(s_i) \neq 0)$ **then**
    | $Sum + = C_j$
    end
  end
  if (*Sum* != 0) **then**
  | $\mathbb{T} = \mathbb{T} \cup s_i$
  end
end
return $\mathbb{T}$

---

section. The Trojan location can be traced by observing the fact that the outputs can possibly be affected by the existing Trojan. We apply the tests, simulate the circuit, and compare the outputs with the golden outputs (golden outputs can be found from the specification polynomials) and keep track of faulty outputs in set $E = \{e_1, e_2, .., e_n\}$. Each e_i denotes one of the erroneous outputs. To localize the Trojan, we partition the gate-level netlist to find fanout-free cones (set of gates that are directly connected together) of the implementation. Each gate whose output is connected to more than one gate is selected as a fanout. For generality, gates that produce primary outputs are also considered as fanouts. To partition the implementation, gate-level netlist as well as a list of fanouts (L_{fo}) are taken into consideration. In each iteration, one fanout-gate is chosen from list L_{fo} and gate-level netlist is traced backward until the gate g_i is reached. The inputs of g_i can come from one of the fanouts in the list L_{fo} or primary inputs. All of the visited gates are marked as one cone. This process continues until all of the fanouts are visited.

Algorithm 4 shows the Trojan localization procedure. Given a partitioned erroneous circuit and a set of faulty outputs E, the goal of the automatic Trojan localization is to identify all of the potentially responsible cones for the Trojan. First, we find a set of cones $C_{e_i} = \{c_1, c_2, \ldots, c_j\}$ that constructs the value of each e_i from set E (lines 4–5). These cones contain suspicious gates. We intersect all of the suspicious cones C_{e_i}s to prune the search space and improve the efficiency of Trojan localization algorithm. The intersection of these cones is stored in C_S (lines 7–8).

Algorithm 4: Trojan localization algorithm

Input: Partitioned Netlist, Faulty Outputs E
Output: Suspected Regions C_S
for *each faulty output $e_i \in E$* **do**
| find cones that construct e_i and put in C_{e_i}
end
$C_S = C_{e_0}$
for $e_i \in E$ **do**
| $C_S = C_S \cap C_{e_i}$
end
return C_S

When the effect of the Trojan can be observed in multiple outputs, it means that the Trojan resides in the intersection of cones which constructs the faulty outputs. We use this information to detect and correct the Trojan. We describe the details of anomaly correction technique in Sect. 4.3.2.

Example 4.5 Consider the faulty 2-bit multiplier shown in Fig. 4.6. Suppose that the AND gate with inputs (M, N) has been replaced with an OR gate by mistake. So, the remainder is $R = 4.A_1.B_0 + 4.A_0.B_1 - 8.A_0.A_1.B_0.B_1$. The assignments that activate the Trojan are calculated based on method demonstrated in Sect. 4.3. Tests are applied and the faulty outputs are obtained as $E = \{Z_2, Z_3\}$. Then, the netlist is partitioned to find fanout-free cones. The cones involved in the construction of faulty outputs are: $C_{Z_2} = \{2, 3, 4, 6, 7\}$ and $C_{Z_3} = \{2, 3, 4, 6, 8\}$. The intersection of the cones that produce faulty outputs is $C_S = \{2, 3, 4, 6\}$. As a result, gates $\{2, 3, 4, 6\}$ are potentially responsible as the source of Trojan. ■

4.3.2   Trojan Correction

After test generation and Trojan localization, the next step is Trojan detection. The remainder is helpful since it contains valuable information about the nature of the Trojan and its location. For example, when the tampered gate is located in the first level (inputs of tampered gates are primary inputs), it creates certain patterns in the remainder. These specific patterns are due to the termination of

Fig. 4.6 Malicious gate-level implementation of a 2-bit multiplier with associated tests

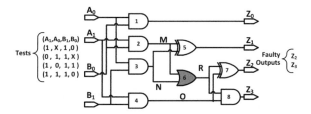

Table 4.2 Remainder patterns caused by gate misplacement Trojan

Suspicious gate	Appeared remainder's pattern	Solution
AND (a,b)	$P_1 : -a-b+2.a.b$	$S_1 :$ OR (a,b)
	$P_2 : -a-b+3.a.b$	$S_2 :$ XOR (a,b)
OR (a,b)	$P_1 :$ a+b$-2.a.b$	$S_1 :$ AND (a,b)
	$P_2 :$ a.b	$S_2 :$ XOR (a,b)
XOR (a,b)	$P_1 :$ a+b$-3.a.b$	$S_1 :$ AND (a,b)
	$P_2 : -a.b$	$S_2 :$ OR (a,b)

the substitution process in the algebraic rewriting after this level, which prevents Trojans from propagating any further. In Example 4.3, the first level XOR gate is placed by mistake instead of an AND gate. Let us consider the effect of a gate-replacement Trojan from algebraic point of view. The equivalent algebraic value of Z_0 is $M = A_0 + B_0 - 2.A_1.B_0$ in the erroneous implementation; however, in the correct implementation, Z_0 should be equal to $Z_0^* = A_0.B_0$. Thus, the difference between Z_0 and Z_0^*, $(A_0 + B_0 - 3.A_1.B_0)$ will be observed in the remainder. Therefore, whenever $a + b - 3.a.b$ pattern is seen in the remainder and there is an XOR gate with inputs (a, b) in the implementation, we can conclude that the XOR gate is the source of Trojan and it should be replaced with an AND gate. Table 4.2 shows the patterns that will be observed for misplacement of different types of gates. Note that gates with three (or more) inputs can be modeled as cascades of 2-input gates. So, the patterns are also valid for complex gates.

From Sect. 4.3.1, we have a set of cones C_S such that their gates are potentially responsible for the trigger of the Trojan. First, the gates in C_S are extracted and they are kept in a set $\mathbb{G}$. Next, the suspicious gates from the first level of $\mathbb{G}$ are considered and the remainder is scanned to check whether one of the patterns in Table 4.2 is recognized. If the pattern is found, the Trojan gate is replaced with the corresponding gate. Otherwise, the terms of the remainder are rewritten such that it contains output variable of first level gates (at this time, we are sure that the first level gates are not the cause of the problem). We also remove the safe gates from $\mathbb{G}$. Then, we repeat the process over the remaining gates in $\mathbb{G}$ until we find the source of the Trojan.

Example 4.6 Consider the untrustworthy circuit shown in Example 4.5. The remainder is $R = 4.A_1.B_0 + 4.A_0.B_1 - 8.A_0.A_1.B_0.B_1$, and the potentially malicious gates are numbered as 2, 3, 4, and 6. As we can see, remainder R does not contain any patterns shown in Table 4.2. It means that the first level suspicious gates 2, 3, and 4 are not responsible for the Trojan. Thus, we try to rewrite the remainder's terms with the output of the correct gates. In this step, we know that gates 2, 3, and 4 are correct so their algebraic expressions are also true. As gate 6 is the only remaining gate, it is the answer. However, we continue the process to show the final solution. By considering $M = A_1.B_0$ and $N = A_0.B_1$, R will be rewritten as $R^* = 4.(M + N - 2.M.N)$ (signal's weight is computed as shown in [20]). Now, we consider the gates in the second level. This time R^* matches

with one of the patterns shown in Table 4.2. Based on Table 4.2, an AND gate with (M, N) as its inputs has been replaced with an OR gate. The only gate that has these characteristics is gate 6 which is also in $\mathbb{G}$. It means that the source of the Trojan has to be the gate 6 and if replaced with an AND gate, the Trojan will be removed.
■

Finding and factorizing of remainder terms in order to rewrite them would be complex for larger designs. To overcome the complexity and obviate the need for manual intervention, we present an automated approach shown in Algorithm 5. The algorithm takes faulty gate-level netlist, remainder R, and potentially malicious gates of set $\mathbb{G}$ (sorted based on their levels) as inputs. It starts from the first level gate g_i; if g_i is the source of the existing malfunction in the design, one of the patterns in Table 4.2 should have been manifested in the remainder based on g_i's type. Therefore, the anomaly correction algorithm computes two patterns (P_1, P_2) with g_i's inputs (lines 7–12) and scans the remainder to check whether one of them matches. If one of the patterns is found, the Trojan is identified and it can be corrected based on Table 4.2 (lines 13–16). Otherwise, g_i is correct and it will be removed from set $\mathbb{G}$ and next gate will be selected. Moreover, the current algebraic expression of g_i is true and it can be used in subsequent iterations (gate g_j from higher levels gets the output of g_i as one of its inputs, the expression of g_i can be used instead of its output variables). Since our goal is to compute patterns such that they contain just primary inputs, we use a dictionary to keep the expression of the gate output based on the primary inputs (line 19). The weight of each gates' output is computed based on the weight of its inputs. The weights of the primary inputs and primary outputs are known a priori. The weights of any internal signals can be computed recursively utilizing forward as well as backward traversal. We can also utilize the following properties for different gates. For XOR and OR gates, the weight of the output is same as inputs weight. In multipliers, the weight of the output of the first level AND gates is computed as multiplication of weights of the inputs (they are responsible for partial products). On the other hand, the weights of the output of other AND gates in the design is computed as the sum of weights of the inputs (since they are mostly used in half-adders [20]). In adders, the weight of the output of all AND gates is computed as union of weights of the inputs. This process continues until the Trojan is detected or set $\mathbb{G}$ is empty. Since, the algorithm starts from primary inputs, it will not reach a gate whose inputs do not exist in the dictionary. Note that the anomaly detection approach does not need all of the counter-examples to work. It works even if there is no counter-example (all of the gates are considered as suspicious) or there is just one counter-example. However, having more counter-examples improves the detection performance.

Example 4.7 We want to apply Algorithm 5 to the case shown in Example 4.6. We start from gate 2 and compute $P_1 = -2.A_1 - 2.B_0 + 4.A_1.B_0$ and $P_2 = -2.A_1 - 2.B_0 + 6.A_1.B_0$ for gate 2. As these patterns do not exist in the remainder, gate 2 is correct and the dictionary will be updated as $(M = 2.A_1.B_0)$. The same will happen for gates 3 and 4, and the dictionary will be updated as $(M = 2.A_1.B_0, N = 2.A_0.B_1)$ at the end of this iteration. When we consider gate 6, the P_is are as follows:

Algorithm 5: Trojan detection/correction

Input: Suspected Gates G, Remainder R
Output: Trojan Gate and Solution
sort g_i based on their levels (lowest level first)
for *each level j* **do**
    for *each $g_i \in G$ from level j* **do**
        $(a, b) = inputs(g_i)$
        if *!(each of (a, b) are from PI)* **then**
           $a = dic.get(a)$
           $b = dic.get(b)$
        end
        $P_1 = Compute\,P_1(a, b)$
        $P_2 = Compute\,P_2(a, b)$
        if *(P_1 is found in R)* **then**
           return *gate g_i and solution S_1 from Table 4.2*
        end
        if *(P_2 is found in R)* **then**
           return *gate g_i and solution S_2 from Table 4.2* **else**
             r
        end
        emove g_i from G
        dic.add(output(g_i), Expression($g_i(a, b)$)))
    end
end

$P_1 = 4.A_1.B_0 + 4.A_0.B_1 - 8.A_1.B_0.A_0.B_1$ and $P_2 = 4.A_1.B_0.A_0.B_1$. Considering that $R = 4.A_1.B_0 + 4.A_0.B_1 - 8.A_0.A_1.B_0.B_1$, P_1 of gate 6 can be observed in R. So the Trojan is the OR gate 6, and based on Table 4.2 it can be fixed by replacing with an AND gate. ∎

Signal inversion problem can be viewed in the same way as gate-replacement Trojan. If we consider a wire as a buffer, it may be replaced with an inverter. Therefore, it is a special case of gate-replacement Trojan, where a buffer can be replaced with an inverter, or vice versa. For example, assume that signal a is inverted by mistake in the actual implementation. Therefore, the difference between the expected behavior and the implementation appears in the remainder by performing the functional rewriting of the specification polynomial. In this case, instead of a we encounter $1-a$ in the implementation, and the remainder is $R = 1-a-a = 1-2*a$. As a result, the appearance of the pattern $1 - 2 * a$ in the remainder reveals the fact that signal a is inverted by mistake.

4.4   Detecting Multiple Trojans

Section 4.3 presented algorithms for detecting, localizing, and correcting a single Trojan. In this section, we extend these algorithms for detecting multiple Trojans.

The threat model (gate replacement) as well as remainder generation process remains the same. If the algebraic rewriting of an arithmetic circuit results in a non-zero remainder, we know that the implementation is untrustworthy. However, the sources of the Trojans are unknown. The plan is to use the non-zero remainder in order to generate directed tests to activate the Trojans, localize the source of Trojans, and correct them. First, we explain how we extend the approach presented in Sect. 4.3 to correct multiple Trojans with independent triggers. Then, we present an approach to automatically detect two Trojans with independent triggers.

If there is more than one Trojan in the implementation, the remainder will be affected by all of them since all of the malicious gates are contributing in the algebraic rewriting procedure as well as the remainder generation. In other words, the remainder shows the effect of all Trojans in the implementation. Example 4.8 shows how the remainder is generated when there are two Trojans in the implementation.

Example 4.8 In the circuit shown in Fig. 4.7, the AND gate with inputs (A_0, B_0) and the AND gate with inputs (A_1, B_1) are replaced with XOR and OR gates, respectively (i.e., two Trojans in the implementation of a 2-bit multiplier). The result of algebraic rewriting (remainder polynomial) can be computed as shown in Eq. 4.4. ∎

$$f_{spec} : 8.Z_3 + 4.Z_2 + 2.Z_1 + Z_0 - 4.A_1.B_1 - 2.A_1 B_0 - 2.A_0.B_1 - A_0.B_0$$

$$step_1 : 4.R + 4.O + 2.z_1 + Z_0 - 4.A_1.B_1 - 2.A_1 B_0 - 2.A_0.B_1 - A_0.B_0$$

$$step_2 : 4.O + 2.M + 2.N + Z_0 - 4.A_1.B_1 - 2.A_1 B_0 - 2.A_0.B_1 - A_0.B_0$$

$$step_3(remainder) : R = A_0 + B_0 - 3.A_0.B_0 + 4.A_1 + 4.B_1 - 8.A_1.B_1$$

$$(4.4)$$

Detailed observation in the remainder generation procedure shows that the overall remainder can be considered as the sum of different individual Trojan's effect in the algebraic rewriting process. For instance, the first part of the remainder shown in Example 4.5 comes from the remainder shown in Example 4.3 (the same

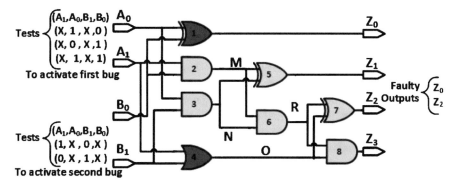

Fig. 4.7 Gate-level netlist of a 2-bit multiplier with two gates replacement (dark gates) as well as associated tests to activate them

Trojan) as $(A_0 + B_0 - 3.A_0.B_0)$, and the second part $(4.A_1 + 4.B_1 - 8.A_1.B_1)$ is responsible for the second Trojan. Clearly, the second part is equal to the remainder that can be the result of the algebraic rewriting with an implementation which contains only the second Trojan. Therefore, each assignment that makes the remainder non-zero activates at least one of the existing malicious scenarios. Some tests may activate all of the Trojans at the same time. Thus, Algorithm 3 can be used to generate directed tests when there are more than one fault in the design.

Example 4.9 Directed tests to activate the malicious implementation of Example 4.8 are shown in Fig. 4.7. The assignments make the first part of the remainder non-zero $(A_0 + B_0 - 3.A_0.B_0)$, and activate the first Trojan. For example, assignment $(A_1 = 1, A_0 = 0, B_1 = 0, B_0 = 0)$ manifests the effect of the first Trojan in Z_0. On the other hand, the assignments that make the second part of the remainder non-zero $(4.A_1 + 4.B_1 - 8.A_1.B_1)$ are tests to activate the second Trojan. Assignment $(A_1 = 1, A_0 = 0, B_1 = 0, B_0 = 0)$ activates the second Trojan in Z_2. However, the assignment $(A_1 = 1, A_0 = 0, B_1 = 0, B_0 = 1)$ activates both of these Trojans at the same time $(Z_0$ and $Z_2)$.   ∎

To localize the source of Trojans, the circuit is simulated using the generated tests to find faulty primary outputs. Malicious gates exist in the cones that construct the functionality of faulty outputs. In order to prune the search space and localize source of Trojans, we cannot directly apply Algorithm 4 as their intersection may be a zero set. However, some information can be found from using Algorithm 4. In the following sections, we describe the Trojan localization and correction of multiple Trojans: (1) Sect. 4.4.1 covers Trojans with independent trigger cones (independent Trojans), and (2) Sect. 4.4.2 covers Trojans which share some trigger cones (dependent Trojans).

4.4.1   Removing Multiple Independent Trojans

We refer two Trojans as independent if they have different trigger cones (fan-ins). Figure 4.7 shows two independent Trojans in a 2-bit multiplier. If multiple Trojans are independent of each other, their effect can be observed easily in the remainder as the sum of each individual Trojan's remainder (sum of sub-remainders). Therefore, if the remainder is partitioned into multiple sub-remainders based on the primary inputs (each part representing the effect of one Trojan), each sub-remainder as well as the associate malicious cones can be fed into Algorithm 5 in order to detect and correct the source of multiple independent Trojans.

If the input cones (input fan-ins) of malicious gates are separate from each other, a different set of primary inputs may appear in each sub-remainder. In order to find the sub-remainders, each term of the overall remainder and its corresponding monomial are examined to determine which sub-remainder it belongs. Algorithm 6 shows the remainder partitioning procedure.

Algorithm 6: Remainder partitioning

Input: Remainder R
Output: Sub-remainders $\mathbb{R}$
Sort terms of R based on their size
$R_0 = largestTerm(R)$
$\mathbb{R} = \{R_0\}$
for *each term* $t \in R$ **do**
    for *each sub-remainder* $R_i \subset \mathbb{R}$ **do**
        if (R_i *contains some of the variable* t) **then**
            $R_i = R_i + t$
        else
            n
        end
        ew $R_j = t$
        $\mathbb{R} = \mathbb{R} \cup R_j$
    end
end
return $\mathbb{R}$

Algorithm 6 takes the overall remainder R as input and returns the partitioned sub-remainders R_is. The algorithm sorts the terms of the R based on their monomial size (the number of variables in each term) in descending order (line 5). In the next step, it starts from the largest term of the remainder R and adds it to sub-remainder R_0 (line 6). Then, it examines all terms of R from the second largest term t to find out which partition they belong to (lines 7–8). If some of the variables of the term t already exist in the sub-remainder R_i, the term t will be added to sub-remainder R_i (lines 9–10). Otherwise, the algorithm creates a new sub-remainder R_j and adds t to it (lines 12–13). The process continues until all terms of the R are examined. If the algorithm results in only one sub-remainder, it shows that malicious gates do not have independent input cones. The computed sub-remainders are fed into Algorithm 3 in order to generate directed tests activating the corresponding Trojan of that sub-remainder. The generated tests are used to define the corresponding faulty outputs of each Trojan. Example 4.10 illustrates the remainder partitioning procedure.

Example 4.10 Consider the faulty multiplier design shown in Fig. 4.7 and corresponding remainder shown in Eq. 4.4. In order to find different possible sub-remainders, the remainder is sorted as: $R = -3.A_0.B_0 - 8.A_1.B_1 + A_0 + B_0 + 4.A_1 + 4.B_1$. The partitioning starts from term $-3.A_0.B_0$ and as there are no sub-remainder so far, sub-remainder R_1 is created and the term is added to it as: $R_1 = -3.A_0.B_0$. The second term $-8.A_1.B_1$ is examined and as R_1 does not contain variables A_1 and B_1, new sub-remainder R_2 is created. Similarly, rest of the terms of R are examined and R_1 and R_2 are computed as: $R_1 = -3.A_0.B_0 + A_0 + B_0$ and $R_2 = -8.A_1.B_1 + 4.A_1 + 4.B_1$. The directed tests are shown in Fig. 4.7.

The generated tests are applied and faulty outputs are defined. The faulty outputs of each Trojan are fed into Algorithm 4 in order to find potential malicious cones.

Algorithm 5 is used with each sub-remainder as well as corresponding potential malicious gates as its inputs, and it tries to detect and correct each Trojan. In other words, the problem of security verification of an untrustworthy design with n independent Trojans is mapped to fixing of n malicious designs where each design contains a single Trojan. We illustrate how to apply Algorithm 5 to remove multiple independent sources of Trojans using Example 4.11.

Example 4.11 Having the directed tests shown in Fig. 4.7, faulty outputs Z_0 and Z_2 as well as two sub-remainders computed in Example 4.10, Algorithm 5 is used twice to find the source of Trojans. In the first attempt, the faulty output is Z_0 and the computed suspicious cone using Algorithm 4 contains only gate 1. In this gate, gate 1 and R_1 are fed into the bug correction algorithm (Algorithm 5). Two patterns $P_1 = A_0 + B_0 - 3.A_0.B_0$ (if the potential malicious gate 1 should be an AND gate) and $P_2 = -1.A_0.B_0$ (if the suspicious gate 1 should be an OR gate) are computed. Therefore, gate 1 should be replaced with an AND gate to fix the first Trojan since the P_1 is equal to the remainder R_1. The same procedure is used for the second Trojan while the potential malicious gates are $\{2, 3, 4, 6, 7\}$ since the only faulty output is Z_2. Trying different patterns results in a conclusion that gate 4 should be replaced with an AND gate.   ■

4.4.2 Removal of Dependent Trojans

In this section, we describe how to detect and correct dependent Trojans that share their triggers' logic. The key difference here from the cases that we solved in Sect. 4.4.1 is the fact that the remainder cannot easily be partitioned into sub-remainders since some of the terms of the corresponding sub-remainder may be canceled through other sub-remainders or they may be combined to each other. The reason is that the Trojans share triggers' logic (fan-ins) and their individual sub-remainders may have common terms consisting of a set of primary inputs as variables. When sub-remainders are combined to each other to form the overall remainder, some term combinations/cancellations happen. Moreover, some of the sub-remainders may be affected by lower level Trojans and the presented method in Sect. 4.4.1 cannot solve these cases. We illustrate the fact using the following example.

Example 4.12 Consider the faulty implementation of a 2-bit multiplier with two Trojans as shown in Fig. 4.8. Assume that gates 6 and 8 are replaced with OR gates to inject Trojans. It can be observed from Fig. 4.8 that two Trojans share some set of input cones (gates $\{2, 3, 4\}$ are common in input cones of Trojan gates 6 and 7). Applying algebraic rewriting on the circuit shown in Fig. 4.8 results in a non-zero remainder: $R = 8.A_1.B_1 + 12.A_1.B_0 + 12.A_0.B_1 - 16.A_0.A_1.B_0 - 16.A_1.B_0.B_1$. However, if only gate 6 is replaced with an OR gate in the implementation (single Trojan), the remainder will be equal to: $R_1 = 4.A_0.B_1 + 4.A_1.B_0 - 8.A_0.A_1.B_0.B_1$. Similarly, when only gate 7 is replaced with an OR gate (single Trojan), the

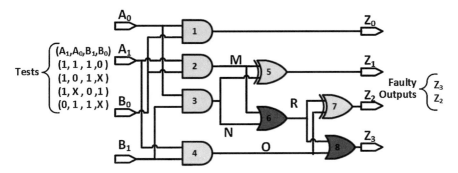

Fig. 4.8 Gate-level netlist of a 2-bit multiplier with two Trojans (dark gates) which shares some input cones as well as associated tests to activate them

remainder will be computed as: $R_2 = 8.A_1.B_1 - 8.A_0.A_1.B_0.B_1$. As it can be observed, $R \neq R_1 + R_2$. The reason is that corrupted gate 6 has an effect on the generation of sub-remainder R_2. As a result, R'_2 should be computed as: $R'_2 = 8.A_1.B_1 + 8.A_0.B_1 + 8.A_0.B_1 - 16.A_0.A_1.B_0 - 16.A_1.B_0.B_1 + 8.A_0.A_1.B_0.B_1$. We can observe that $R = R_1 + R'_2$. Note that there is not any monomial of $A_0.A_1.B_0.B_1$ in the remainder R; however, this monomial exists in both R_1 and R'_2 with opposite coefficients resulting in the term cancellation. ∎

As it can be observed from Example 4.12, term cancellation and lower level Trojans' effect are two main reasons that limit the applicability of the algorithms presented in Sect. 4.4.1 to detect and correct Trojans with common input cones. In this section, we present a general approach to correct and detect multiple gate misplacement Trojans regardless of their positions.

The first step to fix dependent Trojans is to use Algorithm 3 in order to generate directed tests to activate unknown Trojans. In the next step, the circuit is simulated using the generated tests to define the faulty outputs (E) since the effect of Trojans will be propagated to them. Algorithm 4 cannot be used to localize the potential malicious cones since the intersection of the malicious cones may eliminate some of the malicious gates. Instead, union of all of the gates that construct faulty outputs should be considered as suspicious gate candidates to make sure that all of Trojan candidates are considered. The next step is to define malicious gates and their corresponding solutions using the remainder as well as potential malicious gates. We construct two sub-remainders from each potentially malicious gates (e.g., considering if the current gate is suspicious and the type of gate is AND, the solution can be either OR gate or XOR gate based on Table 4.2) and we store them in set $\mathbb{R}$. To be able to detect the Trojans, we are looking for n sub-remainders $R_i \in \mathbb{R}$ where their union constructs the original remainder R.

In general, finding n dependent Trojans and constructing the respective remainder R map to "subset sum" problem and it has exponential complexity. In other words, we need to find n potential sub-remainders such that their sum is equal to the remainder R. Therefore, for each gate in a malicious region, we construct two patterns (sub-remainders) as shown in Algorithm 5 as total m sub-remainders. To be able to detect and correct n dependent Trojans, we need to select $\binom{m}{n}$, where $r_1 + r_2 + \ldots + r_n = R$. The most naïve algorithm to solve this problem is to consider all subset of m sub-remainders, and check whether the subset sums to R for every subset. The complexity of this algorithm is in the order of $O(2^m)$. If we use the naïve approach for finding two dependent Trojans, the complexity is $O(m^2)$. By introducing Algorithm 7 and using a hash map, we could solve this problem in linear time $O(m)$ for two dependent Trojans.

Figure 4.1 shows all malicious scenarios that the current method can automatically detect in linear time. As it can be seen in Fig. 4.1d, the method is capable of handling $2 * k$ dependent Trojans in linear time when we have k independent malicious regions where each of them has at most two dependent Trojans. In other words, Algorithm 6 partitions the remainder R into k sub-remainders where for each sub-remainder r_i, the method tries to find at most two dependent Trojans in linear time.

To detect two dependent Trojans in a malicious region, we are looking for two sub-remainders such that their sum constructs the overall remainder R. Note that sub-remainder of an individual Trojan may be affected by the other existing Trojan in the implementation (for instance, sub-remainder R'_2 which shows the effect of Trojan gate 7 in Example 4.12 is also affected by Trojan gate 6). Algorithm 7 corrects two dependent Trojans by finding two sub-remainders R_1 and R_2 such that their sum is equal to R ($R = R_1 + R_2$). The algorithm tries to find two equal polynomials: $R - R1$ and $R2$. The algorithm takes the remainder and potential malicious gates as inputs and it returns two malicious gates and their correct replacement as output. The algorithm consists of three major steps. First, polynomials corresponding to gate's inputs (we have assumed that a gate has two inputs for simplicity in representation) are computed based on primary inputs for each potentially malicious gate g_i (a and b are corresponding polynomials of gate g_i inputs). Computed polynomials are added to map dic (lines 6–8). Second, algorithm constructs two patterns (P_1 and P_2) for each suspicious gates g_i based on Table 4.2 regarding the functionality of their input gates (lines 9–12). Note that P_1 and P_2 can be constructed based on the fact that we have considered only three types of gates (AND, OR, and XOR) so that each gate can be replaced by two other ones. For example, if the suspicious gate is an AND gate, it can be replaced with either an OR gate or an XOR gate to fix the Trojan. Therefore, we construct two patterns, one showing the functionality of replacing the AND gate with an OR gate (P_1), and the other one shows the functionality of replacing the AND gate with an XOR gate (P_2). Computed patterns are added to set $\mathbb{P}$ (line 13). For computed patterns P_1 and P_2, the algorithm computes the $R - P_i$ and it stores the result in a map $\mathbb{R}$ (lines 14–15). In the final step, each of the patterns $P_j \in \mathbb{P}$ is checked to see whether it exists in the map $\mathbb{R}$ (lines 16–18). If P_j exists in map $\mathbb{R}$, it means that there were a pattern P_i in set $\mathbb{P}$, where $R - P_i = P_j$. Therefore, P_i and P_j are the sub-remainders R_1 and R_2 that we are looking for such that $R_1 = P_i$ and $R_2 = P_j$. The gates corresponding to P_i and P_j are the source of Trojans and their solution can be found based on Table 4.2 (lines 19–20). Note that by using hash map $\mathbb{R}$

Algorithm 7: Fixing two dependent Trojans

Input: Suspicious gates $\mathbb{G}$, remainder R
Output: malicious gates and their solution
$\mathbb{P} = \{\}$ /*A set that keeps patterns for all gates as well as corresponding solution of each pattern*/
$\mathbb{R} = \{\}$ /*A map that keeps remainder minus all patterns as well as corresponding patterns*/
for *each gate $g_i \in \mathbb{G}$* **do**
$\quad$ $(a, b) = \text{computeInputPolynomials}(g_i)$
$\quad$ dic.add(g,(a, b)))
end
for *each gate $g_i \in \mathbb{G}$* **do**
$\quad$ $(a, b) = \text{dic.getInputPolynomials}(g_i)$
$\quad$ $P_1 = compute P1(a, b)$
$\quad$ $P_2 = compute P2(a, b)$
$\quad$ $\mathbb{P} = \mathbb{P} \cup \{P_1, P_2\}$
$\quad$ $\mathbb{R}.put((R - P_1), P_1)$
$\quad$ $\mathbb{R}.put((R - P_2), P_2)$
end
for *each $P_j \in \mathbb{P}$* **do**
$\quad$ **if** *P_j exists in $\mathbb{R}$* **then**
$\quad\quad$ $P_i = \mathbb{R}.get(P_j)$
$\quad\quad$ gate $g_i = \mathbb{P}.get(P_i)$ is malicious and get solution S_i from Table 4.2
$\quad\quad$ gate $g_j = \mathbb{P}.get(P_j)$ is malicious and get solution S_j from Table 4.2
$\quad$ **end**
end

---

the complexity of the algorithm is proportional to the number of malicious gates. The complexity of the algorithm grows linearly with the number of suspicious gates (suspicious gates can be obtained by Trojan localization phase).

Note that Algorithm 7 requires to construct the exact sub-remainder responsible for the potential Trojans (it is not useful to find the pattern as some part of the remainder). The exact sub-remainder is dependent on the gates that the corrupted gates are connected in the next level of the design. To illustrate the point, suppose that gate g_1 is connected to only a half-adder with inputs g_1 and g_2. If f_{g_1} and f_{g_2} show the corresponding polynomials of gates g_1 and g_2 based on the functionality of their inputs, gate g_1 contributes to the functionality of the next level by polynomial

$$f_{g_1} + f_{g_2} - 2 * f_{g_1} * f_{g-2}(\text{XOR}) + 2 * f_{g_1} * f_{g-2}(\text{AND}) = f_{g_1} + f_{g_2}$$

However, if the gate g_1 is the source of the Trojan and its functionality is replaced by polynomial $f_{g_1'}$, there would be a difference in the functionality of the design as: $\Delta = f_{g_1'} - f_{g_1}$. If gate g_1 is connected to a half-adder with inputs g_1 and g_2, the reduction results in $\Delta + f_{g_2} - 2.\Delta.f_{g_2} + 2.\Delta.f_{g_2}$. Since f_{g_2} should be included in the correct functionality of the design, the exact sub-remainder can be computed as: $\Delta - 2.\Delta.f_{g_2} + 2.\Delta.f_{g_2} = \Delta$. Patterns that are computed in Table 4.2 match with Δ based on the polynomials of inputs of the malicious gate. In arithmetic circuit implementations, most of the gates are connected to half-adders (or they are in the last level of the design). Therefore, if we consider them as Trojan candidates, their constructed patterns are equal to the exact remainder.

However, if suspicious gate g_1 is not connected to a half-adder, the exact sub-remainder due to malicious gate g_1 may include more terms besides the terms in Δ. For example, if g_1 is connected to an XOR gate g_2, the exact remainder would be equal to: $\Delta - 2.\Delta.f_{g_2}$. The extra part $-2.\Delta.f_{g_2}$ comes from vanishing monomial propagated to the remainder due to the effect of the Trojan and no counterpart monomials will appear to cancel them during backward algebraic rewriting.

Example 4.13 Consider the faulty full-adder shown in Fig. 4.9. The gate G_2 has been replaced by an OR gate to inject a Trojan in the gate-level implementation. After the verification procedure, the remainder is: $R = 2 * (A + B - 2 * A * B) - 2 * C_{in} * (A + B - 2 * A * B)$. The remainder R has two parts: the first part shows the difference of the functionality of the malicious gate (OR) and the correct gate (AND) as: $\Delta = (A + B - A * B) - A * B = A + B - 2 * A * B$. However, the second part $(-2 * C_{in} * (A + B - 2 * A * B))$ represents the vanishing monomials propagated to the remainder due to the Trojan.

In order to construct the exact remainder for a suspicious gate g_i, we construct Δ patterns based on Table 4.2. In the second step, we consider each gate g_j such that g_i is its input and we compute the corresponding polynomial g_j based on its inputs' polynomials. The terms that contain Δ should be added to the remainder. Note that, if we have two cascaded Trojans, the effect mentioned above only happens for the higher level Trojan since the effect of the lower level Trojan is considered while constructing the pattern of the higher level Trojan. Another important aspect is that the weight of each gate should be considered as we described in Sect. 4.3.2.

Example 4.14 Consider the faulty full-adder shown in Fig. 4.9 where gate G_2 has been replaced by an OR gate (it should be an AND gate in the correct implementation). We know that the remainder is equal to $R = 2 * ((A + B - 2 * A * B) - C_{in} * (A + B - 2 * A * B))$ and the implementation is untrustworthy. If we are suspicious about the G_2 and we guess that it should be an AND gate, we construct $\Delta = A + B - 2 * A * B$ based on Table 4.2. Since G_2 is only input of gate G_3, we construct the polynomial as

Fig. 4.9 Faulty netlist with one Trojan (gate G_2 should have been an AND gate)

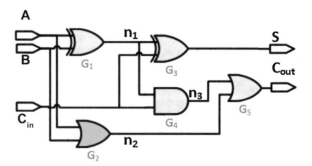

$$f_{G_4} + \Delta - \Delta * f_{G_4}$$

Since the term f_{G_4} is not dependent on Δ, it is a part of the correct functionality of the implementation, and it should not be considered in the remainder. Therefore the constructed remainder is

$$R' = \Delta - \Delta * f_{G_4} = 2 * ((A + B - 2 * A * B) - C_{in} * (A + B - 2 * A * B))$$

As $R = R'$, we can conclude that gate G_2 is corrupted, and it should be replaced by an AND gate to fix the Trojan. We show that how exact sub-remainders are used to detect two dependent Trojans in Example 4.15.

Example 4.15 Consider the faulty implementation of a 2-bit multiplier shown in Fig. 4.8 with remainder: $R = 8.A_1.B_1 + 12.A_1.B_0 + 12.A_0.B_1 - 16.A_0.A_1.B_0 - 16.A_1.B_0.B_1$. Corresponding directed tests to activate existing Trojans are shown in Fig. 4.8. Trojan candidates are computed based on faulty outputs Z_2 and Z_3 as gates $\{2, 3, 4, 6, 7, 8\}$. Algorithm 7 creates two patterns for each of the suspicious gates as shown in Table 4.3 column "Pattern." For each pattern, the possible solution and remainder minus patterns are listed in the third and fourth columns of Table 4.3, respectively. Note that Table 4.3 is the combination of two lists, $\mathbb{P}$ and hash map $\mathbb{R}$, which are mentioned in Algorithm 7. Each pattern listed in the second column is tested to find whether it exists in hash map $\mathbb{R}$ (part of hash map is shown in the fourth column). As it can be seen in the table, P_{11} (highlighted polynomial in the second column) is equal to $R - P_7$ (highlighted in the fourth column). It means that $R - P_7 = P_{11} \rightarrow R = P_{11} + P_7$. Therefore, gates 6 and 8 are malicious since P_7 and P_{11} are corresponding to these gates and they should be substituted with AND gates. ∎

4.5 Challenges in Remainder Generation

Depending on the location of the Trojan, the remainder generation can be challenging. However, this detection approach works as long as the remainder exists. The number of terms drastically grows when the bug is in the deeper stages of the design. Therefore, it is essential to provide a mechanism that generates efficient and compact remainders.

The reason for complexity growth is that the malicious gate may introduce new terms during the intermediate steps of the specification polynomial's reduction. These extra terms are multiplied to polynomials of other gates and grow continuously until the remainder contains only primary inputs (we call it remainder's terms explosion effect). Theoretically, a remainder can contain 2^n terms, where n is the number of primary inputs. The challenges of remainder generation may limit the applicability of using symbolic algebra for security verification of Trojan-inserted design, especially when the Trojan is deep inside the design. There are

Table 4.3 Patterns for suspicious gates in Example 4.15

$Gate\#$	Pattern	Solution	Remainder minus pattern
2 (AND)	$2.A_1 + 2.B_0 - 4.A_1.B_0$	OR	$-2.A_1 - 2.B_0 + 8.A_1.B_1 + 16.A_1.B_0 + 12.A_0.B_1 - 16.A_0.A_1.B_0 - 16.A_1.B_0.B_1$
	$2.A_1 + 2.B_0 - 6.A_1.B_0$	XOR	$-2.A_1 - 2.B_0 + 8.A_1.B_1 + 18.A_1.B_0 + 12.A_0.B_1 - 16.A_0.A_1.B_0 - 16.A_1.B_0.B_1$
3 (AND)	$2.A_0 + 2.B_1 - 4.A_0.B_1$	OR	$-2.A_0 - 2.B_1 + 8.A_1.B_1 + 12.A_1.B_0 + 16.A_0.B_1 - 16.A_0.A_1.B_0 - 16.A_1.B_0.B_1$
	$2.A_0 + 2.B_1 - 6.A_0.B_1$	XOR	$-2.A_0 - 2.B_1 + 8.A_1.B_1 + 12.A_1.B_0 + 18.A_0.B_1 - 16.A_0.A_1.B_0 - 16.A_1.B_0.B_1$
4 (AND)	$4.A_1 + 4.B_1 - 8.A_1.B_1$	OR	$-4.A_1 - 4.B_1 + 16.A_1.B_1 + 12.A_1.B_0 + 12.A_0.B_1 - 16.A_0.A_1.B_0 - 16.A_1.B_0.B_1$
	$4.A_1 + 4.B_1 - 12.A_1.B_1$	XOR	$-4.A_1 - 4.B_1 + 20.A_1.B_1 + 12.A_1.B_0 + 12.A_0.B_1 - 16.A_0.A_1.B_0 - 16.A_1.B_0.B_1$
6 (OR)	$4.A_0.B_1 + 4.A_1.B_0 - 8.A_0.A_1.B_0.B_1$	AND	$8.A_1.B_1 + 8.A_1.B_0 + 8.A_0.B_1 - 16.A_0.A_1.B_0 - 16.A_1.B_0.B_1 + 8.A_0.A_1.B_0.B_1$
	$4.A_0.A_1.B_0.B_1$	XOR	$8.A_1.B_1 + 12.A_1.B_0 + 12.A_0.B_1 - 16.A_0.A_1.B_0 - 16.A_1.B_0.B_1 + 4.A_0.A_1.B_0.B_1$
7 (XOR)	$4.A_0.B_1 + 4.A_1.B_0 + 4.A_1.B_1 - 8.A_0.A_1.B_0 - 8.A_1.B_0.B_1 + 4.A_0.A_1.B_0.B_1$	AND	$4.A_1.B_1 + 8.A_1.B_0 + 8.A_0.B_1 - 8.A_0.A_1.B_0 - 8.A_1.B_0.B_1 + 8.A_0.A_1.B_0.B_1$
	$4.A_0.A_1.B_0 + 4.A_1.B_0.B_1 - 4.A_0.A_1.B_0.B_1$	OR	$8.A_1.B_1 + 12.A_1.B_0 + 12.A_0.B_1 - 20.A_0.A_1.B_0 - 20.A_1.B_0.B_1 + 4.A_0.A_1.B_0.B_1$
8 (OR)	$8.A_1.B_1 + 8.A_0.B_1 + 8.A_0.B_1 - 16.A_0.A_1.B_0 - 16.A_1.B_0.B_1 + 8.A_0.A_1.B_0.B_1$	AND	$8.A_1.B_1 + 12.A_1.B_0 + 12.A_0.B_1 - 16.A_0.A_1.B_0 - 16.A_1.B_0.B_1$
	$8.A_0.A_1.B_0 + 8.A_1.B_0.B_1 - 8.A_0.A_1.B_0.B_1$	XOR	$8.A_1.B_1 + 12.A_1.B_0 + 12.A_0.B_1 - 24.A_0.A_1.B_0 - 24.A_1.B_0.B_1 + 8.A_0.A_1.B_0.B_1$

approaches such as [17] to address the aforementioned challenge by generating more compact remainders. This approach expedites the remainder generation time, and it also reduces the number of terms in the remainder and makes it possible to generate a remainder irrespective of the location of the Trojan. In other words, this approach helps in remainder's term explosion effect. More compact remainders can be generated based on partitioning the input space of the design. The presented approach is based on applying certain constraints on primary inputs and solve the verification problem for each input constraint. If set $\mathbb{M} = \{0, 1\}^n$ shows all input combinations of a design with input bits $\{x_0, x_1, \ldots, x_{n-1}\}$ and if specification ($\mathbb{S}$) and implementation ($\mathbb{I}$) are equivalent for all combinations of ($\mathbb{S} \overset{\mathbb{M}}{\equiv} \mathbb{I}$), they should also be equivalent for any input combinations that belong to $\mathbb{M}$ ($\forall M \subset \mathbb{M}, \quad \mathbb{S} \overset{M}{\equiv} \mathbb{I}$). If the implementation is Trojan-inserted, at least one of the intermediate reductions will result in a non-zero remainder. Algorithm 8 shows the input partitioning approach. Given the set of primary inputs K with a particular order, the algorithm returns n different constraints on primary inputs, where n is the number of primary inputs. Initially, the algorithm sets all of the inputs to zero except the first input in set K which is kept in the symbolic form, and the algorithm adds them to the set of results $\mathbb{M}$ (lines 5–8). In the next step, it keeps the first input in the symbolic form and sets the second input of the ordered set as "1," and sets other inputs to "0," and adds the constraints to the result. This process continues until all of the inputs are kept in their symbolic form except the last one which is set to true. The variable *index* presents the index of primary inputs that should be assigned to true (line 11). The variables before the index variable are kept in their symbolic form, and variables that come after the index are assigned to false (lines 12–15). In each iteration, the *index* variable is updated (line 16). The algorithm returns the set of constraints as output. This algorithm guarantees (see the proof in [17]) that the entire inputs' space is covered since all of the combinations of primary inputs are considered (each input bit is assigned to either one, zero, or kept in the symbolic form which can take both values).

Algorithm 8: Generation of input constraints

Input: Primary inputs K
Output: Set of Constraints Map $\mathbb{M}$
new map $M = \{\}; n = sizeOf(K)$
$M.add(0, K[0])$
for $i = 1; i < n; j++$ **do**
  | $M.add(i, false)$
end
$\mathbb{M}.add(M), index = 1$
for $i = 0; i \leq n; i++$ **do**
  | $M = \{\}$
  | $M.put(index, true)$
  | **for** $j = 0; j < index; j++$ **do**
  | | $M.add(j, K[j])$
  | **end**
  | **for** $j = index + 1; i \leq n; j++$ **do**
  | | $M.put(j, false)$
  | **end**
  | $index++$
  | $\mathbb{M}.add(M)$
end
return $\mathbb{M}$

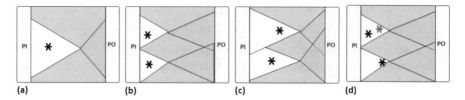

Fig. 4.10 Faulty netlist with one tampered gate (gate 8 should have been an AND)

Table 4.4 Input constraints to efficiently verify the untrustworthy circuit shown in Fig. 4.10	A_1	B_1	A_0	B_0
	A_1	0	0	0
	A_1	1	0	0
	A_1	B_1	1	0
	A_1	B_1	A_0	1

Example 4.16 Assume that we want to partition the input space of the 2-bit multiplier shown in Fig. 4.10 using Algorithm 8. Suppose that primary inputs are given in the following order: $\{A_1, B_1, A_0, B_0\}$. Table 4.4 shows the four different constraints on primary inputs. It can be easily verified that these four constraints cover the entire primary inputs' space. The first and second rows cover two combinations each, the third row covers four combinations, and the last row covers eight combinations. Therefore, the four input constraints in Table 4.4 can cover all sixteen combinations.   ∎

In this chapter, we present an incremental remainder generation method using the constraints computed based on Algorithm 8. The original verification problem is mapped to n sub-problems where the specification and implementation polynomials are updated by applying the corresponding constraints. In each sub-problem, a new set of implementation polynomials is computed based on propagating the integer values of the corresponding constraint and considering them while constructing polynomials of each gate and each fanout-free region. Specification polynomial is also updated by applying the conditions of primary inputs in the original specification polynomial. In each sub-problem, the corresponding specification polynomial is reduced over the related implementation polynomials. If the remainder is non-zero, the given constraint manifests some Trojans in the design. The implementation and specification of an arithmetic circuit are equivalent if remainders of each of the n sub-problems is computed as a zero remainder.

Example 4.17 Consider the 2-bit multiplier shown in Fig. 4.10. We want to apply the incremental equivalence checking approach of Algorithm 9 using all of the input constraints shown in Table 4.4 to verify the correctness of the implementation. Equation 4.5 shows the steps of the verification. Specification and implementation polynomials are updated using each constraint. For instance, polynomial of gate 3 is computed as: $N = A_0 * B_1 = 0$ as A_0 and B_1 are considered zero in the first iteration

---

Algorithm 9: Incremental remainder generation algorithm

Input: Input constraint M_i, specification polynomial f_{spec}, Gate-level netlist C
Output: Remainder r if the implementation is malicious
for *each input constraints $M_i \in \mathbb{M}$* **do**
$\quad f_{spec_i}$ =findSpecificationPolynomial(f_{spec}, M_i)
$\quad \mathbb{F}_i$ =findImplementationPolynomials(C, M_i)
$\quad r_i$ = reduction of f_{spec_i} over $f_j s \in \mathbb{F}_i$
$\quad$ **if** $(r_i! = 0)$ **then**
$\quad\quad$ Implementation is Trojan-inserted
$\quad\quad$ **return** r_i
$\quad$ **end**
end
return *0* /*correct implementation for constraint M_i */

(first row of Table 4.4). Since the last iteration generates a non-zero remainder, the implementation is malicious. ∎

$$\mathbb{F}_1 = \{Z_0 = 0, M = 0, N = 0, O = 0, R = 0, Z_1 = 0, Z_2 = 0, Z_3 = 0\}$$

$$f_{spec_1} : 8 * Z_3 + 4 * Z_2 + 2 * Z_1 + Z_0$$

$$step_{11}(remainder) : 0$$

$$\mathbb{F}_2 = \{Z_0 = 0, M = 0, N = 0, O = A_1, R = 0, Z_1 = 0, Z_2 = A_1, Z_3 = A_1\}$$

$$f_{spec_2} : 8 * Z_3 + 4 * Z_2 + 2 * Z_1 + Z_0 - 4 * A_1$$

$$step_{12} : 2 * Z_1 + Z_0 + 8 * A_1$$

$$step_{22}(remainder) : 8 * A_1$$

$$(4.5)$$

4.5.1 Ordering of Primary Inputs

Ordering of primary inputs to produce inputs' constraints impacts the performance of the incremental security verification approach when the implementation is Trojan-inserted. The size of the remainder depends on the location of the Trojans. The Trojan in the deeper stages of the design causes a larger remainder such that generation of the remainder is impossible due to the remainder's terms explosion effect in traditional equivalence checking approaches. The remainder grows through the procedure of reduction of f_{spec} over implementation polynomials. The very first time that the functionality of the tampered gate is involved in the intermediate steps of reduction of f_{spec}, the core of the remainder is formed by terms showing the difference of the malicious functionality from the expected functionality (δ). Terms of the remainder grow gradually during the reduction by substituting terms of the δ with the functionality of gates in the input cone of the malicious cone. Therefore, this approach is extremely helpful while verifying integer arithmetic circuits which contain long carry chain and the functionality of primary outputs is dependent on earlier stages of the design. Therefore, the incremental equivalence checking is more beneficial when it verifies the functionality of deeper stages of the design while deactivating lower stage of the design in order to reduce the problem complexity and reducing terms of the potential remainder. To make this approach applicable for any Trojan in the design, we explore incremental remainder generation based on dynamic ordering using binary search. We start from a predefined order of input constraints (as shown in Table 4.4) and try the first row, if the first row generates a non-zero remainder, we are done. We have the remainder and we can start the security verification. Otherwise, we divide the table of the constraints into two and select the first row in the second half of the table. If the constraints of this row lead to a non-zero remainder, we continue dividing until we reach to (1) the last row of the table and we still have a zero remainder, (2) to a row which generates a non-zero remainder, (3) a row that using its constraints in remainder generation leads to a number of terms which is higher than a predefined threshold. In the first case, we can conclude that the design is correct since the last row has all of the variables in the symbolic form. In the second case, the remainder is generated and we can use it for Trojan detection. In the last case, the remainder cannot be generated due to term explosion effect. In fact, all of the rows below this row will face the same problem since the number of symbolic variables increases as we move towards the bottom of the table. In this case, the upper rows may generate a more compact remainder. The key observation is that from the last row that generates a zero remainder till the row that has a term explosion effect, some new input conditions have triggered the Trojan. If we give those input conditions higher priority, the remainder will be generated in a more effective way.

4.6 Experiments

4.6.1 Experimental Setup

The directed test generation, Trojan localization, and Trojan removal algorithms were implemented in a Java program and experiments were conducted on a Windows PC with Intel Xeon Processor and 16 GB memory. We have tested this approach on both pre- [13] and post-synthesized gate-level arithmetic circuits that implement adders and multipliers. Post-synthesized designs were obtained by synthesizing the high-level description of arithmetic circuits using Xilinx synthesis tool. We consider wrong gate (gate-replacement Trojan) or signal inversion which changes the functionality of the design as the threat model. Several gates from different levels were replaced with an erroneous gate in order to generate faulty implementations. The remainders were generated based on the method presented in [17]. Multiple counter-examples (directed tests) are generated based on one remainder. As each counter-example can be generated independent of others, so we used a parallelized version of the algorithm for faster test generation. We compared the test generation method with existing directed test generation method [8] as well as random test generation. Several Trojans are inserted in the middle levels of the circuits to conduct this experimental results. We compared the detection/correction results with most recent work in this context [20]. We use the benchmarks obtained from the authors [20]. However, we have implemented their algorithm to compare this method with their method. To enable fair comparison, similar to [20], we randomly inserted Trojans (gate changes) in the middle stages of the circuits. We improved the runtime complexity of presented method in [16] by using efficient data structures such as hash maps and sorted sets.

4.6.2 Detecting and Fixing a Single Trojan

Table 4.5 presents results for test generation, Trojan localization, and Trojan removal methods using multipliers and adders. The first column ("Type") indicates the types of benchmarks. The second ("Size") and third ("#Gates") columns show the size of operands and number of gates in each design, respectively. Since the sizes of adder designs are smaller than multiplier designs, we show results only for higher operand sizes (bit-widths). The fourth column ("RG (s)") shows the CPU time to generate the remainder. The fifth column ("Dir. [8] (s)") indicates results for directed test generation method presented in [8] by using SMV model checker [7] (we give the model checker the advantage of knowing the Trojan). The sixth column ("Random (s)") represents results of random test generation method (time to generate the first counter-example using the random technique). The seventh column ("Pro. TG (s)") represents the time of the test generation method that generates multiple tests. As it can be observed from Table 4.5, this method has improved directed test generation time by several orders of magnitude. The eighth column ("Trojan Loc. (s)") shows the CPU time for Trojan localization algorithm. The ninth column ("[20] (s)") shows the detection/correction time of [20] using the implementation of this approach in Java. The next column ("Pro. (TG+BL+DC) (s)") provides CPU time of the presented approach which is the sum of test generation (TG), Trojan localization (BL) and detection/correction (DC) time. The last column ("Improvement") shows the improvement provided by the presented framework. Clearly, this approach is an order of magnitude faster than the most closely related approach [20], especially for larger designs as Trojan localization has an important effect. The reported numbers are the average of generated results for several different scenarios. For instance, if we zoom into test generation of the first row (post-synthesized multiplier with 4-bit operands) of Table 4.5, the reported results are the average of the nine possible scenarios shown in Table 4.6.

Table 4.6 presents the anomaly detection/correction results of 4-bit post-synthesized multiplier. The first column ("Trojans") shows a possible set of gate-replacement Trojans. Time to generate the first counter-example using model checker [8] and random techniques is reported in the second ("Dir. [8] (s)") and

Table 4.5 CPU time and memory results of security verification of arithmetic circuits

Benchmark			Verification	Test generation (TG)			Trojan localization (BL)	Detection/Correction (DC)		
Type	Size	#Gates	RG (s)	Dir. [8] (s)	Random (s)	Pro. TG (s)	Trojan Loc. (s)	[20] (s)	Pro. (RG+TG+BL+DC) (s)	Improv.
post-syn. Multipliers	4	72	0.07	1.88	0.02	0.01	0.001	0.2	0.09	2.22×
	16	1632	1.26	42.69	1.48	0.32	0.03	4.32	2.04	2.11×
	32	6848	3.57	205.66	3.03	0.82	0.16	18.50	5.97	3.1×
	64	28K	14.31	MO	16.97	1.65	0.83	151.05	27.94	5.41×
	128	132K	58.64	MO	66.52	3.83	5.1	1796.50	111.55	16.10×
	256	640K	319.75	MO	TO	15.65	22.39	TO	524.76	–
pre-syn. Multipliers	4	94	0.05	1.27	0.04	0.01	0.001	0.17	0.08	2.13×
	16	1860	1.45	43.11	1.93	0.4	0.03	4.45	2.28	1.95×
	32	7812	3.61	189.50	5.69	0.87	0.2	23.1	6.28	3.68×
	64	32K	12.36	MO	29.07	1.77	0.8	180.3	27.27	6.61×
	128	129K	50.60	MO	83.60	4.1	3.8	1743.07	98.34	17.72×
	256	521K	225.72	MO	TO	12.44	15.83	TO	396.2	–
post-syn. Adder	64	573	0.50	154.97	1.51	0.5	0.01	3.12	1.21	2.58×
	128	1251	1.04	MO	3.48	1.07	0.05	6.60	2.73	2.41×
	256	2301	3.52	MO	10.64	3.09	0.05	17.32	7.79	2.22×
pre-syn. Adder	64	444	0.58	128.12	1.15	0.35	0.01	2.95	1.09	2.71×
	128	880	1.08	MO	4.40	0.84	0.03	6.46	2.2	2.93×
	256	1698	3.9	MO	9.10	2.23	0.1	16.18	7.44	2.17×

Single Trojan inserted in the middle stages of the design as well as close to primary inputs. TO = timeout after 3600 s; MO = memory out of 8 GB

Table 4.6 Test generation time for 4-bit multiplier with 8 bits outputs # Gates = 72

Trojans	Dir. [8] (s)	Ran. (s)	#Tests	Faulty outputs	# Ran. tests	Pro. TG (s)
XOR → AND	1.48	0.05	18	Z_7, Z_6, Z_5, Z_4	2632	0.01
XOR → OR	2.12	0.03	4	Z_2	2945	0.01
XOR → AND	1.95	0.02	128	Z_4	2292	0.01
XOR → OR	2.27	0.03	12	Z_6, Z_5, Z_4, Z_3	2945	0.05
XOR → AND	1.03	0.02	14	Z_6, Z_5, Z_4, Z_3, Z_2	2369	0.02
AND → XOR	2.44	0.05	3	Z_6, Z_5, Z_4, Z_3, Z_2	1881	0.01
AND → OR	2.20	0.002	2	Z_7, Z_6, Z_5	2258	0.01
AND → XOR	0.89	0.04	148	Z_7, Z_6, Z_5, Z_4	2164	0.03
OR → AND	2.52	0.01	148	Z_6	2920	0.01
Average	1.88	0.03	53	–	2489.55	0.01

third columns ("Ran. (s)"), respectively. The fourth column ("#Tests") shows the number of directed tests generated by this approach to activate the Trojan (each of them activates the Trojan). The fifth column ("Faulty Outputs") lists the outputs that are affected by the Trojan (activated by the respective tests reported in the "#Tests" column). The sixth column ("#Ran. Tests") shows the number of random tests required to cover all of the directed tests. It demonstrates that even for such small circuits, using random tests to activate the Trojan is impractical. The last column ("Pro. TG (s)") shows the test generation time. As mentioned earlier, the average of these scenarios is reported in the first row of Table 4.6.

The experimental results demonstrate three important aspects of this approach. First, the test generation method generates multiple directed tests when the Trojan is unknown in a cost-effective way. Second, this approach detects and corrects single Trojan caused by gate replacement in a reasonable time. Finally, this method is not dependent on any specific architecture of arithmetic circuits and it can be applied on both pre-synthesized and post-synthesized gate-level circuits.

4.6.3 Detection/Correction of Multiple Trojans

Table 4.7 presents results for remainder generation, remainder partitioning, test generation, Trojan localization, and detection/correction methods using multipliers and adders with multiple independent Trojans. The first column ("Type") indicates the types of benchmarks. The second ("Size") and third columns ("#Trojans") show the size of operands and number of Trojans in each design, respectively. The fourth column ("RG (s)") shows the CPU time to generate the remainder. The fifth column ("RP (s)") represents the required time for remainder partitioning, and the sixth column ("TG (s)") represents the time of the test generation method. The seventh column ("Trojan Loc. (s)") shows the CPU time for Trojan localization algorithm. The eighth column ("DC (s)") shows the detection/correction time to detect and

Table 4.7 CPU time and memory results for security verification of arithmetic circuits for multiple independent Trojans

Type	Size	#Trojans	RG (s)	RP (s)	TG (s)	Trojan Loc. (s)	DC (s)	Total (RG+RP+TG+BL+DC) (s)	[20] (s)	Imp.	Mem
post_syn. multipliers	8 × 8	4	0.15	0.001	0.04	0.03	0.47	0.69	1.7	2.46×	6.4 MB
		8	0.21	0.001	0.07	0.03	0.76	1.08	2.5	2.31×	7.6 MB
	16 × 16	4	1.37	0.003	0.57	0.01	1.22	3.17	6.03	1.90×	29.53 MB
		8	1.41	0.003	1.2	0.02	1.62	4.25	10.07	2.36×	31.88 MB
	32 × 32	4	3.68	0.003	1.86	0.64	4.52	10.43	26.37	2.53×	48.00 MB
		8	3.73	0.003	2.08	1.18	8.4	15.4	43.98	2.85×	58.65 MB
	64 × 64	4	14.87	0.006	5.65	3.9	31.48	55.9	178.89	3.20×	76.3 MB
		8	15.1	0.006	7.06	4.7	45.31	73.17	250.07	3.41×	102.1 MB
	128 × 128	4	58.9	0.008	11.59	10.1	114.52	195.12	1946.1	9.97×	378.5
		8	67.25	0.008	25.67	20.87	175.88	289.68	2337.56	8.07×	406.3 MB
	256 × 256	4	356.10	0.012	39.58	70.65	508.42	983.76	TO	–	1.38 GB
		8	372.8	0.012	65.21	122.01	706.22	1266.25	TO	–	1.65 GB
pre_syn. multipliers	8 × 8	4	0.21	0.001	0.44	0.03	0.25	0.93	1.73	1.86×	1.95 MB
		8	0.22	0.001	0.5	0.03	0.46	1.21	2.67	2.21×	2.14 MB
	16 × 16	4	1.48	0.002	1.3	0.05	1.5	4.33	7.4	1.68×	7.24 MB
		8	1.55	0.002	1.90	0.05	1.87	5.34	10.05	1.88×	8.1 MB
	32 × 32	4	3.9	0.003	2.08	0.73	5.8	12.51	30.34	2.42×	23.07 MB
		8	3.97	0.003	3.23	1.31	7.98	16.49	43.18	2.62×	30.56 MB
	64 × 64	4	13.71	0.001	5.94	4.5	33.22	57.37	194	3.38×	97.40 MB
		8	14.09	0.005	7.91	6.9	69.5	112.42	225.85	2.01×	103.2 MB
	128 × 128	4	53.42	0.006	13.5	15.09	170.46	252.47	2036.37	8.06×	222.46 MB
		8	68.64	0.006	22.48	26.72	207.88	352.73	2260.6	6.4×	250.6 MB
	256 × 256	4	283.97	0.01	26.75	39.16	653	1002.89	TO	–	0.92 GB
		8	294.75	0.01	59.13	77.34	866.18	1297.41	TO	–	0.97 GB

post_syn. adders	64 × 64	4	0.51	0.004	0.89	0.07	0.22	1.7	3.43	2.02×	3.2 MB
		8	0.52	0.003	1.63	0.12	0.27	2.56	3.85	1.50×	3.25 MB
	128 × 128	4	1.07	0.005	3.14	0.2	0.71	5.12	7.59	1.48×	5.5 MB
		8	0.97	0.01	5.41	0.37	0.85	7.81	8.72	1.12×	5.62 MB
	256 × 256	4	3.67	0.01	8	0.3	3.62	15.6	19.87	1.27×	12.3 MB
		8	3.71	0.01	11.44	0.8	5.94	21.9	25.93	1.18×	14.1 MB
pre_syn. adders	64 × 64	4	0.40	0.002	0.88	0.08	0.3	1.76	3.30	1.90×	2.8 MB
		8	0.41	0.006	1.12	0.08	0.32	1.94	3.56	1.84×	3.5 MB
	128 × 128	4	1.1	0.008	3.19	0.2	0.79	5.29	7.26	1.65×	6.9 MB
		8	1.32	0.009	3.57	0.42	1.01	6.32	8.31	1.37×	6.98 MB
	256 × 256	4	3.54	0.01	6.24	0.28	4.38	14.45	19.13	1.33×	13.1 MB
		8	3.6	0.01	9.52	0.81	5.27	19.21	23.35	1.21×	14.3 MB

TO = timeout after 7200 s. Trojans are inserted in the middle stages of the design as well as close to primary inputs correct all Trojans. The next column ("total (RP+TG+BL+DC) (s)") provides CPU time of the presented approach which is the sum of remainder partitioning (RP), test generation (TG), Trojan localization (BL), and Trojan correction algorithm (DC) times. The tenth column ("[20] (s)") shows the required time of the method presented in [20] using the implementation of this approach in Java. The next column ("Improvement") shows the improvement provided by this framework. Clearly, this approach is an order of magnitude faster than the most closely related approach [20], especially for larger multipliers as Trojan localization has an important effect. However, the performance is comparable with [20] for security verification of adders since the number of gates is small and the number of inputs is large and test generation time may surpass the speed up of this method. The last column shows the required memory for the whole approach.

Table 4.8 presents results for remainder partitioning, test generation, Trojan localization, and detection/correction methods using multipliers and adders with two dependent Trojans. The first column ("Type") indicates the types of benchmarks. The second column ("Size") shows the size of operands. The third column ("RG(s)") shows th CPU time to generate the remainder. The fourth column ("RP (s)") represents the required time for remainder partitioning, and the fifth column ("TG (s)") represents the time of the test generation method. The sixth ("BL (s)") and seventh ("DC (s)") columns show the CPU time for Trojan localization and detection/correction time, respectively. Trojan localization time is relatively small

Table 4.8 CPU time and memory results for security verification of arithmetic circuits with two dependent Trojans

Type	Size	RG (s)	RP (s)	TG (s)	BL (s)	DC (s)	Total (s)	Mem
post_syn. mul.	8	0.18	0.001	0.1	0.01	0.98	1.09	11.72 MB
	16	1.43	0.002	0.35	0.02	2.23	4.04	40.97 MB
	32	3.45	0.002	0.96	0.08	13.92	18.39	60.21 MB
	64	14.3	0.004	3.77	0.2	77.12	95.4	83.3 MB
	128	54.22	0.008	8.06	0.6	241.05	303.93	365 MB
	256	310.13	0.012	31.8	36.02	1099.96	1477.92	1.36 GB
pre_syn. mul.	8	0.21	0.001	0.1	0.01	0.91	1.23	8.8 MB
	16	1.52	0.001	0.77	0.01	5	7.3	22.4 MB
	32	3.88	0.002	1.03	0.08	13.54	18.53	50.32 B
	64	13.82	0.003	4.65	0.1	96.3	114.87	79.2 MB
	128	59.13	0.005	7.88	0.6	220.22	287.83	293 MB
	256	280.04	0.01	19.41	22.05	982.9	1584.45	1.02 GB
post_syn. mul.	64	0.5	0.001	01.18	0.01	0.55	2.24	3.3 MB
	128	1.1	0.011	5.4	0.02	3.47	10	7.01 MB
	256	3.7	0.011	16.09	0.1	9.42	29.32	11.96 MB
pre_syn. add.	64	0.4	0.003	1.13	0.01	0.53	2.07	2.9 MB
	128	1.21	0.008	6.3	0.01	2.36	9.89	8.2 MBB
	256	3.5	0.01	10.97	0.08	15.04	27.6	14.01 MB

Trojans are inserted in the middle stages of the design as well as close to primary inputs in comparison with other scenarios since the intersection of malicious cones is not computed. The next column ("Total (s)") provides CPU time of the presented approach which is the sum of remainder partitioning (RP), test generation (TG), Trojan localization (BL), and Trojan detection/correction algorithm (DC) times. As the result shows, this approach can detect and correct multiple dependent Trojans in reasonable time. We did not compare with any approaches since there are no existing approaches for detecting/fixing multiple dependent Trojans. Finally, the last column shows the required memory for security verification approach.

4.7 Summary

In this chapter, we presented an automated methodology for security verification of arithmetic circuits. The methodology consists of efficient directed test generation, Trojan localization, and Trojan correction algorithms. The presented framework used the remainder produced by equivalence checking methods to generate directed tests that are guaranteed to activate the source of the malicious functionality in the design. This approach used the generated tests to localize the source of the Trojan and find suspicious areas in the design. We also presented an efficient security verification algorithm that uses the remainder as well as suspicious areas to locate and correct the Trojan without any manual intervention. We showed that this approach can be extended to automatically fix multiple Trojans. The experimental results demonstrated the effectiveness of the presented approach to find anomalies in large and complex arithmetic circuits in an efficient manner.

References

1. A. Ahmed, P. Mishra, QUEBS: qualifying event based search in concolic testing for validation of RTL models, in *IEEE International Conference on Computer Design (ICCD)* (2017), pp. 185–192
2. A. Ahmed, F. Farahmandi, P. Mishra, Directed test generation using concolic testing of RTL models, in *Design Automation and Test in Europe (DATE)*, pp. 1538–1543 (2018)
3. E. Biham, Y. Carmeli, A. Shamir, Bug attacks, in *Advances in Cryptology* (2008), pp. 221–240
4. R.E. Bryant, Y.-A. Chen, Verification of arithmetic circuits with binary moment diagrams, in *Proceedings of the 32nd Annual ACM/IEEE Design Automation Conference* (ACM, New York, 1995), pp. 535–541
5. B. Buchberger, Some properties of gröbner-bases for polynomial ideals. ACM SIGSAM Bull. **10**(4), 19–24 (1976)
6. B. Buchberger, A criterion for detecting unnecessary reductions in the construction of a Göbner bases, in *EUROSAM* (1979)
7. Cadence Berkeley Lab, The Cadence SMV Model Checker. Available at http://www.kenmcmil.com
8. M. Chen, P. Mishra, Functional test generation using efficient property clustering and learning techniques. IEEE Trans. Comput. Aided Des. Integr. Circuits Syst. **29**(3), 396–404 (2010)

9. M. Chen, P. Mishra, Property learning techniques for efficient generation of directed tests. IEEE Trans. Comput. **60**(6), 852–864 (2011)
10. M. Chen, X. Qin, H. Koo, P. Mishra, *System-level Validation – High-Level Modeling and Directed Test Generation Techniques* (Springer, New York, 2012)
11. M. Chen, P. Mishra, D. Kalita, Automatic RTL test generation from SystemC TLM specifications. ACM Trans. Embed. Comput. Syst. **11**(2), article 38 (2012)
12. M. Chen, X. Qin, P. Mishra, Learning-oriented property decomposition for automated generation of directed tests. J. Electron. Test. **30**(3), 287–306 (2014)
13. M.J. Ciesielski, C. Yu, W. Brown, D. Liu, A. Rossi, Verification of gate-level arithmetic circuits by function extraction, in *IEEE/ACM International Conference on Computer Design Automation(DAC)* (2015), pp. 1–6
14. D. Cox, J. Little, D. O'shea, *Ideals, Varieties, and Algorithms*, vol. 3 (Springer, New York, 1992)
15. F. Farahmandi, B. Alizadeh, Gröbner basis based formal verification of large arithmetic circuits using Gaussian elimination and cone-based polynomial extraction, in *Microprocessors and Microsystems – Embedded Hardware Design* (2015), pp. 83–96
16. F. Farahmandi, P. Mishra, Automated test generation for debugging arithmetic circuits, in *2016 Design, Automation & Test in Europe Conference & Exhibition (DATE)* (IEEE, Piscataway, 2016), pp. 1351–1356
17. F. Farahmandi, P. Mishra, Automated debugging of arithmetic circuits using incremental Gröbner basis reduction, in *2017 IEEE 35th International Conference on Computer Design (ICCD)* (IEEE, Piscataway, 2017), pp. 193–200
18. F. Farahmandi, P. Mishra, Automated test generation for debugging multiple bugs in arithmetic circuits. IEEE Trans. Comput. **68**(2), 182–197 (2019)
19. F. Farahmandi, B. Alizadeh, Z. Navabi, Effective combination of algebraic techniques and decision diagrams to formally verify large arithmetic circuits, in *2014 IEEE Computer Society Annual Symposium on VLSI* (IEEE, Piscataway, 2014), pp. 338–343
20. S. Ghandali, C. Yu, D. Liu, W. Brown, M. Ciesielski, Logic debugging of arithmetic circuits, in *2015 IEEE Computer Society Annual Symposium on VLSI* (IEEE, Piscataway, 2015), pp. 113–118
21. G.-M. Greuel, G. Pfister, H. Schifinemann, SINGULAR 3.1.3 A computer algebra system for polynomial computations. Centre for Computer Algebra (2012). http://www.singular.uni-kl.de
22. M. Knežević, K. Sakiyama, J. Fan, I. Verbauwhed, Modular reduction in $GF(2^n)$ without precomputational phase, in *Proceedings of the International Workshop on Arithmetic of Finite Fields* (2008), pp. 77–87
23. C. Koc, T. Acar, Montgomery multiplication in $GF(2^k)$, in *Designs, Codes and Cryptography*, vol. 14 (1998), pp. 57–69
24. H.-M. Koo, P. Mishra, Functional test generation using design and property decomposition techniques. ACM Trans. Embed. Comput. Syst. **8**(4), article 32 (2009)
25. B. Le, H. Mangassarian, B. Keng, A. Veneris, Non-solution implications using reverse domination in a modern SAT-based debugging environment, in *Design Automation and Test in Europe (DATE)* (2012), pp. 629–634
26. J. Lv, P. Kalla, F. Enescu, Efficient Gröbner basis reductions for formal verification of Galois field multipliers, in *Proceedings Design, Automation and Test in Europe Conference (DATE)* (2012), pp. 899–904
27. J. Lv, P. Kalla, F. Enescu, Efficient Gröbner basis reductions for formal verification of Galois field arithmetic circuits. IEEE Trans. CAD **32**, 1409–1420 (2013)
28. Y. Lyu, X. Qin, M. Chen, P. Mishra, Directed test generation for validation of cache coherence protocols. IEEE Trans. Comput. Aided Des. Integr. Circuits Syst. **38**, 163–176 (2018)
29. Y. Lyu, A. Ahmed, P. Mishra, Automated activation of multiple targets in RTL models using concolic testing, in *Design Automation and Test in Europe (DATE)* (2019)
30. H. Mangassarian, A. Veneris, S. Safarpour, M. Benedetti, D. Smith, A performance-driven QBF-based iterative logic array representation with applications to verification, debug and test, in *2007 IEEE/ACM International Conference on Computer-Aided Design* (IEEE, Piscataway, 2007), pp. 240–245

31. P. Mishra, N. Dutt, Specification-driven directed test generation for validation of pipelined processors. ACM Trans. Des. Autom. Electron. Syst. **13**(2), 36 pp., article 42 (2008)

32. X. Qin, P. Mishra, Directed test generation for validation of multicore architectures. ACM Trans. Des. Autom. Electron. Syst. **17**(3), article 24, 21 pp. (2012)

33. X. Qin, P. Mishra, Scalable test generation by interleaving concrete and symbolic execution, in *International Conference on VLSI Design* (2014), pp. 104–109

34. O. Wienand, M. Welder, D. Stoffel, W. Kunz, G.M. Greuel, An algebraic approach for proving data correctness in arithmetic data paths, in *Computer Aided Verification (CAV)* (2008), pp. 473–486

Chapter 5
Trojan Localization Using Symbolic Algebra

5.1 Introduction

Intellectual property (IP) outsourcing is a widely used practice in System-on-Chip (SoC) design methodology to reduce the time to market and overall cost. However, it raises major security risks as the attacker can embed malicious components in third-party IPs. Such malicious components, widely known as hardware Trojans, may affect the correct behavior and defeat the trustworthiness of the design by leaking protected information such as secret keys. Hardware Trojans consist of two parts: a trigger and a payload. The trigger is a set of conditions such that their activation deviates the desired functionality from the specification and their effects are propagated through the payload. The adversary designs trigger conditions such that they are satisfied in very rare situations and usually after long hours of operation [3]. Conventional structural and functional testing methods are not effective to activate trigger conditions since there are many possible Trojans and it is not feasible to construct a fault model for each of them. As a result, existing EDA tools are incapable of detecting hardware Trojans and differentiating between trustworthy third-party IPs and untrustworthy ones.

There has been a lot of research on hardware Trojan detection using logic testing and side-channel analysis [1, 3, 7–11]. Logic testing focuses on generating efficient tests to activate a Trojan and check the primary output values of specification and circuit under test to detect Trojan. Side-channel analysis focuses on the difference of the side-channel signature between the golden circuit and Trojan infected circuit. These two types of methods answer the question of whether a circuit is infected with Trojan, but they cannot identify the location of the Trojan. Approaches based on structural/functional analysis [2, 6, 13, 19] have been proposed to identify/localize the malicious logic. Unused circuit identification (UCI) [6] finds for unused portions in the circuit and flags them as malicious. Sturton et al. show that many other types of malicious circuits can evade the detection of the UCI algorithm [15]. The FANCI approach [19] was proposed to flag suspicious nodes based on the concept of control

© Springer Nature Switzerland AG 2020
F. Farahmandi et al., *System-on-Chip Security*,
https://doi.org/10.1007/978-3-030-30596-3_5

values. However, FANCI flags about 1–8% of all nodes, which might be too many suspicious candidates for experts to analyze for a large circuit. Moreover, FANCI returns a set of suspicious nodes even when the circuit is Trojan free. A recent work by Oya et al. [13] manually crafted templates for Trojans and was successful in using these templates to identify Trojans in Trust-HUB benchmarks [18]. Unfortunately, this approach is applicable only for specific types of Trojans; therefore, it is not suitable to detect other types of Trojans that are not covered by their templates. Side-channel analysis focuses on the side-channel signatures of the circuit [1, 9], which avoids the limitations (low trigger probability and propagation of payload) of logic testing. However, the abnormality in side-channel signatures for Trojan circuit is sensitive to measurement noise and process variation, which makes side-channel analysis not effective on large circuits. Narasimhan et al. [12] proposed the temporal self-referencing approach on large sequential circuits. Recently, Yuanwen et al. [7] proposed the multiple excitation of rare switching (MERS) approach to combine the advantages of logic testing and side-channel analysis.

In this chapter, we propose an automated approach to identify untrustworthy IPs and localize malicious functional modifications (if any). The technique is based on extracting polynomials from gate-level implementation of the untrustworthy IP and comparing them with specification polynomials. The proposed approach is applicable when the specification is available. This approach is also useful when a golden design has gone through non-functional transformations such as synthesis, and we would like to ensure that the modified design is trustworthy. This approach is scalable due to manipulation of polynomials instead of BDD-based analysis used in traditional equivalence checking techniques. Experimental results using Trust-HUB benchmarks demonstrate that this approach improves both localization and test generation efficiency by several orders of magnitude compared to the state-of-the-art Trojan detection techniques.

Figure 5.1 presents the overview of the proposed methodology. We extract a set of polynomials from the specification ($\mathbb{S}$). We also derive a set of polynomials ($\mathbb{I}$) from the implementation. Finally, we check the equivalence between two sets $\mathbb{S}$ and $\mathbb{I}$ based on Gröbner basis reduction. Each of the polynomials from the specification, f_{spec_i}, is reduced over a set of corresponding implementation polynomials $\mathbb{I}$ and a set of remainders $\mathbb{R}$ is generated. From symbolic computer algebra, it is known that when $r_i = 0$, gates in Rg (set of gates that contribute in reduction of polynomial f_{spec_i} is called region Rg) have successfully implemented f_{spec_i} and it guarantees that all gates in Rg are safe [5]. Any $(r_i \neq 0) \in \mathbb{R}$ shows a suspicious functionality in the corresponding region Rg and all of the gates in Rg are suspicious candidates. The malicious nodes can be pruned by removing the safe gates from the suspicious candidates. When all of r_i's are equal to zero, the implementation is Trojan free. The proposed method can recognize the Trojan-free implementation from the Trojan-inserted one. This method reports a few gates to indicate the presence of a malicious activity (change of functionality) in the implementation. Since the number of malicious gates is very small, this approach is amenable for an exhaustive test generation to activate the Trojan. This method is applied on Trust-HUB benchmarks

Fig. 5.1 The proposed
hardware Trojan localization
flow

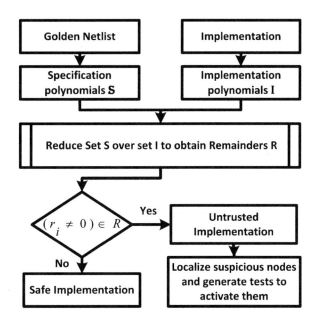

[18] and the experimental results show the effectiveness of the proposed approach compared to existing methods.

The remainder of this chapter is organized as follows: Sect. 5.2 discusses the framework for hardware Trojan localization and detection. Section 5.3 presents the experimental results. Finally, Sect. 5.4 concludes this chapter.

5.2   Trojan Detection and Localization

In order to trust an IP block, we have to make sure that the IP is performing exactly the expected functionality. The approach presented in Sect. 4.2.2 can be extended to find whether a hardware Trojan, which changes the functionality, has been inserted in a combinational arithmetic circuit. However, applying the same approach on general IPs is limited due to several reasons. First, it is possible that the specification of a general circuit cannot be described as one simple polynomial. Second, the circuit may not be acyclic and loops may exist due to their sequential nature. Third, unrolling may increase the complexity of the problem so the reduction of f_{spec} over implementation polynomials will face polynomial terms explosion. Finally, the Trojan activation may require extremely large number of unrolling steps which may be practically infeasible and also there is no specific information on after how many cycles Trojan will be activated. In order to address these challenges, we present a method to generate polynomials in an efficient way and use them in the proposed algorithm to localize and detect Trojans in third-party IPs. The reminder of this section describes the three important tasks in the framework: polynomial generation, Trojan localization, and test generation for Trojan detection.

5.2.1   Polynomial Generation

Suppose that we have two versions of a design, one is a verified IP (specification) and the other is an untrusted third-party IP (implementation) after performing non-functional transformations. The goal is to detect whether an adversary has inserted hard-to-detect hardware Trojan during non-functional changes and has made undesired functional changes. For example, a design house may send their RTL design for synthesis or adding low-power features to a third-party vendor. Once the third-party IP comes back (after synthesis or other functionality-preserving transformations), it is crucial to ensure the trustworthiness of these IPs.

In the method presented in Sect. 4.2.2, specification is modeled as one polynomial; however, here we generate a set of polynomials $\mathbb{S}$ representing the functionality of the golden IP to be able to apply Gröbner basis theory for hardware Trojan localization problem. The specification is partitioned into several regions and each region is converted to a polynomial. The output of each region is either inputs of a flip-flop (clock, enable, reset, etc.) or one of the primary outputs. The inputs of a region are either from primary inputs or inputs/outputs of flip-flops. In other words, we generate polynomials for regions which are limited to flip-flops' boundaries. Then, corresponding equations (based on Eq. 4.1 in Sect. 4.2.2) of gates inside a region are combined together to construct one polynomial representing the functionality of the region.

Algorithm 10: Polynomial generation algorithm

Input: Circuit Graph Gr, L_{out} and L_{in}
Output: Polynomials $\mathbb{S}$
Region = { }
for *each gate* $g_i \in Gr$ *where its output* $\in L_{out}$ **do**
   | Region.add(g_i)
   | **for** *all inputs* g_j *of* g_i **do**
   | | **if** $!(g_j \in L_{in})$ **then**
   | | | Region.add(g_j)
   | | | Call recursively for inputs of g_j over Gr
   | | **end**
   | **end**
   | f_i = convertToPolynomial(Region)
   | $\mathbb{S} = \mathbb{S} \cup f_i$
   | Region = { }
end
return $\mathbb{S}$

Algorithm 10 shows how we extract set $\mathbb{S}$. The specification is converted to a graph where each vertex is a gate (g_i). The algorithm takes the circuit graph Gr, list (L_{out}) of allowed output variables (flip-flops' inputs and primary outputs), and list (L_{in}) of allowed input variables of a region as inputs and returns a set of polynomials $\mathbb{S}$ as its output. The algorithm chooses a gate for which output belongs to L_{out} and goes backward recursively until it reaches the gate g_j, whose input comes from one of the variables from L_{in} (lines 5–10). The algorithm marks all the visited gates as a "Region." The selected region may contain all of the basic gates except flip-flops. Then, the "Region" is converted to a polynomial f_i by combining corresponding polynomials of the gates residing in "Region," f_i is added to set $\mathbb{S}$ (lines 11–12).

Example 5.1 Suppose that the circuit shown in Fig. 5.2 is a part of a verified IP block and we want to use it as specification. Algorithm 10 is applied on it and the polynomials are shown as: $\mathbb{S} = \{f_{spec_1} : n_1 - (-2.A.n_2 + n_2 + A), f_{spec_2} : Z - (1 - n_1.B)\}$. Since the circuit shown in Fig. 5.2 contains one primary output and one flip-flop, Algorithm 10 extracts two specification polynomials for this circuit.

Similarly, the implementation polynomials $\mathbb{I}$ are driven by modeling every gate except flip-flops from the untrusted design as a polynomial based on Eq. 4.1 from Sect. 4.2.2 and Algorithm 10. In order to reduce the number of generated implementation polynomials, we partition implementation to fanout-free cones (set of gates that are directly connected together) and convert each fanout-free region as one polynomial. In other words, $\mathbb{I}$ contains a set of polynomials where each polynomial represents a fanout-free cone.

Example 5.2 The circuit shown in Fig. 5.3 is the Trojan-inserted implantation of the specification shown in Fig. 5.2 (gate 6 is the Trojan trigger and gate 7 is the payload). Gates in same pattern belong to a common fanout-free cone. As a result, set $\mathbb{I}$ is computed by Algorithm 10. Each of the polynomials corresponds to one fanout-free cone.

$$\mathbb{I} = \{n_1 - (n_2.w_4.A - n_2.w_4 + w_4 - n_2.A + n_2),$$
$$w_4 - (A - n_2.A), \tag{5.1}$$
$$Z - (n_1.w_4.C.B - n_1.w_4.C - n_1.B + 1)\}$$

Fig. 5.2 A part of a sequential circuit

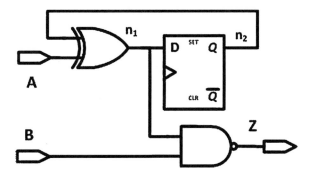

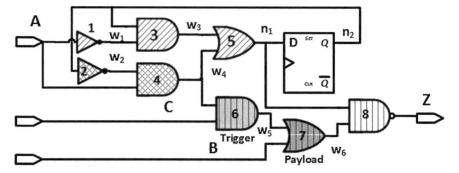

Fig. 5.3 A Trojan-inserted implementation of circuit in Fig. 5.2

5.2.2   Trojan Localization

We generate the set $\mathbb{S}$ and $\mathbb{I}$ as described in Sect. 5.2.1. We assume that the name of flip-flops, primary inputs, and primary outputs is the same between implementation and specification or the name mapping can be done. We also assume that no re-timing has been performed. These are valid assumptions in many scenarios involving third-party IPs. The equivalence of two sets $\mathbb{S}$ and $\mathbb{I}$ is checked to find any suspicious functionality which may serve as a Trojan.

To detect a Trojan, we need to reduce each polynomial f_{spec_i} from set $\mathbb{S}$ over a subset of polynomials from set $\mathbb{I}$ to check membership of every polynomial f_{spec_i} in Ideal I constructed from polynomials from set $\mathbb{I}$ ($I = < \mathbb{I} >$). To perform that, all of the polynomials from $\mathbb{I}$ are hashed based on their leading terms (which contains a single variable and this variable represents the output of the corresponding gate). Every variable from $f_{spec_i} \in \mathbb{S}$ is replaced with the corresponding functionality of that variable from $\mathbb{I}$ polynomials. The process continues until f_{spec_i} is reduced either to zero polynomial or a remainder polynomial which contains primary inputs as well as flip-flop's inputs/outputs. The non-zero remainder indicates that implementation does not correctly implement the functionality of f_{spec_i} and that part of the implementation is suspicious. Note that, based on Gröbner basis theory, when the remainder is zero for a specific region, we can be certain that the region is safe. In other words, it is not possible for a smart attacker to insert malicious gates in a way that the remainder becomes zero.

Example 5.3 Consider we want to measure the trust in the circuit shown in Fig. 5.3, which is the untrustworthy implementation of design shown in Fig. 5.2. Specification polynomials shown in Example 5.1 are reduced over implementation polynomials as shown in Eq. 5.1. The result of the reduction is stored in set $\mathbb{R}$. Each f_{spec_i} produces one remainder r_i that can be either zero or a non-zero polynomial. Gates $\{1, 2, 3, 4, 5\}$ implement functionality of an XOR gate (these gates are equivalent to XOR gate shown in Fig. 5.2). Thus, the remainder r_1 is zero and it means that the region containing gates $\{1, 2, 3, 4, 5\}$ implements the f_{spec_1}

correctly. However, the non-zero remainder r_2 presents the fact that there are malicious components in implementation of f_{spec_2} and the region containing gates $\{2, 4, 6, 7, 8\}$ is suspicious.

$$f_{spec_1} : n_1 + 2.A.n_2 - n_2 - A$$
$$step_{11} : n_2.w_4.A - n_2.w_4 + w_4 + n_2.A - A$$
$$step_{12}(r_1) : 0$$
$$f_{spec_2} : Z + n_1.B - 1 \tag{5.2}$$
$$step_{21} : n_1.w_4.C.B - n_1.w_4.C$$
$$step_{22}(r_2) : -1.n_1.A.C + n_1.n_2.A.C + A.B.C.n_1 - A.B.C.n_1.n_2$$

By using the proposed approach, a set of malicious regions are identified. Suppose the adversary inserts some extra flip-flops as part of Trojans. These buggy flip-flops do not have any correspondence in the specification. In other words, there is no f_{spec_i} which describes their inputs' functionality. Therefore, the corresponding region in the implementation is also considered as a suspicious region. However, scan-chain flip-flops can easily be detected and removed from suspicious candidates because of their structures.

The proposed method formally identifies the regions (between flip-flops boundaries) of the implementation that are safe and the regions that have suspicious functionality. The adversary usually inserts the Trojan in deep levels of the circuit. Therefore, the regions that actually contain the Trojan can be very large and may include many gates (order of hundreds or thousands of gates). In order to improve the approach further, we propose an algorithm to identify the gates that most likely are responsible for the malicious activity. Since we know which regions are Trojan free (based on remainder as zero), we remove the gates which are contributing in the construction of these regions from suspicious regions. In other words, we have formally proved that some of the regions are trustworthy so the gates that construct these regions are essential for the correct functionality. The safe gates may be inputs of Trigger or payload gates. However, they do not belong to the set of malicious gates. Using this approach, we are able to prune the suspicious regions to contain very small number of gates. This approach guarantees that all of the Trojan trigger and payload's gates are inside the suspicious region. Algorithm 11 shows the proposed procedure.

The algorithm takes the gate-level implementation graph G_r as well as specification and implementation polynomials as inputs, and in case the implementation contains malicious components, it returns a set of suspicious gates as output. The algorithm takes each of specification polynomials and reduces them one by one over corresponding polynomials from set $\mathbb{I}$. Each f_{spec_i} may be reduced using several gates g_j and the result of the reduction is stored in r_i (lines 4–5). The used gates are marked to keep track of the gates that are utilized to implement the circuit (line 6). If r_i is equal to zero, it means that all of the g_is are safe and they are

---
Algorithm 11: Hardware Trojan localization algorithm

Input: Circuit implementation G_r, $\mathbb{I}$ and $\mathbb{S}$
Output: Suspicious gates G_t
for *each* $f_{spec_i} \in \mathbb{S}$ **do**
$\quad\quad r_i =$ reduction of f_{spec_i} over $f_j s \in \mathbb{I}$
$\quad\quad R_i = R_i \cup all\ g_j s\ where\ f_j = func(g_j)$
$\quad\quad$ mark all g_is as used
$\quad\quad$ **if** $(r_i! = 0)$ **then**
$\quad\quad\quad\quad R_{TrjIn} = R_{TrjIn} \cup R_i$
$\quad\quad\quad\quad$ **else**
$\quad\quad\quad\quad\quad\quad R_{TrjFree} = R_{TrjFree} \cup R_i$
$\quad\quad\quad\quad$ **end**
$\quad\quad$ **end**
end
for *each gate* $g \in R_{Trjfree}$ **do**
$\quad\quad$ remove g from R_{TrjIn}
end
return $G_t = remaining\ in\ R_{TrjIn} \cup unused\ gates$

---

stored as safe gates ($R_{TrjFree}$); otherwise, all g_is are stored as suspicious candidates (lines 7–11). Every $r_i = 0$ shows that all of the gates used in the construction of functionality of the corresponding f_{spec_i} are safe. Therefore, to narrow down the potential suspicious gates, the gates of G_r which appeared in $R_{TrjFree}$ are removed from R_{TrjIn} (lines 12–13). Note that gates in both of $R_{TrjFree}$ and R_{TrjIn} belong to the implementation G_r. All of the unused gates should also be considered as malicious candidates. Therefore, the union of the remaining gates in R_{TrjIn} and unused gates is returned as potentially malicious gates (G_t). If all of the r_is are zero, the implementation is safe and there is no Trojan inside the implementation.

Algorithm 11 identifies the trust level of a third-party IP and in case of existence of hardware Trojan, it returns a very small number of gates as suspicious candidates. This algorithm guarantees that all of the actual Trojan trigger and payload gates are inside the set G_t.

Example 5.4 Applying Algorithm 11 on the circuit shown in Fig. 5.3 will result in non-zero remainder for region containing gates $\{2, 4, 6, 7, 8\}$. However, the zero remainder of f_{spec_1} shows that gates $\{1, 2, 3, 4, 5\}$ are safe and they are vital to construct the functionality of signal n_1. Therefore, we remove gates $\{2, 4\}$ from potential candidates and gates $\{6, 7, 8\}$ remain as suspicious.

5.2.3   Trojan Activation

As shown in Example 5.4, the small suspicious region still contains some safe gates which are dedicated to the correct functionality in the absence of the Trojan (in Example 5.4, gate 8 is benign but it is reported as suspicious node). In other words, these safe gates are only used to construct the functionality of one specific primary output or flip-flop's input. Thus, they will not be removed in the process of pruning safe gates from suspicious regions since they are not contributing in functionality of other primary outputs or flip-flop's inputs. To be able to detect the exact gates which are responsible for trigger and payload parts of Trojan, we generate tests to activate the Trojan. Since the number of suspicious gates is small enough, we try to activate each node in the suspicious gates and check whether the generated test activates the Trojan. We use an ATPG to generate the directed tests. If none of the tests detects the Trojan, we generate test to activate two of the nodes at the same time. We continue the process until one of the tests activates the Trojan. The proposed method is shown in Algorithm 12. This approach is feasible due to the fact that the number of suspicious nodes that are reported using the proposed approach is very small.

Algorithm 12: Test generation algorithm

Input: Suspicious gates G_t, Implementation C, Specification S
Output: Test vectors T
T={ }
for *each possible trigger scenario n over G_t* **do**
    generate test t_i to activate n of nodes
    for *each possible payload scenario* **do**
        propagate effect of t_i to the observable points
        if *trigger scenario is satisfied* **then**
            $T = T \cup t_i$
        end
    end
end
return T

Example 5.5 We are trying to activate the Trojan shown in Fig. 5.3. From Example 5.4, we know that gates {6, 7, 8} are suspicious. As shown in Fig. 5.3, Trojan will be triggered when output of gate 6 (w_5) becomes true and B is zero at the same time. In other words, gate 8 of the implementation receives one as its second input (w_6), while in the specification, the second input of the NAND gate receives zero. These conditions cause difference between specification and implementation. To propagate the effect of Trojan's condition activation, n_1 should be one since $n_1 = 0$ makes output $Z = 1$ independent of second input's value and it will mask the Trojan effect. The test vectors that activate Trojan are as follows (we assume the initial value of n_2 is equal to 0): $A = 1, B = 0, C = 1$.

5.3   Experiments

5.3.1   Experimental Setup

The Trojan localization algorithm was implemented in a Java program and experiments were conducted on PC with Intel Processor E5-1620 v3 and 16 GB memory. We have tested the approach using widely used Trust-HUB benchmarks [18] consisting of combinational and sequential Trojan triggers and payloads that change the functionality of the design. The Trojan-free designs are considered as specification. To show that this methodology is orthogonal to design structures and library format, we synthesized Trojan-inserted benchmarks with Xilinx synthesis tool and used them as implementation (we just map flip-flops' inputs/output names). Specification is partitioned into several regions and each region is represented using one polynomial. These polynomials can be reduced over implementation polynomials independently. Therefore, we used a parallel version of Algorithm 11 to implement the method. We also used logic reduction based rewriting schemes presented in [14] to improve the equivalence checking time. We compared the results with most relevant Trojan localization work [19]. Since this approach essentially performs equivalence checking, we also compared with an equivalence checking tool "Formality" [16] which has been designed to check the equivalence between two versions of a design to demonstrate the efficiency of this work. When the designers make non-functional changes in a design, Formality tries to detect potential functional changes between two versions of a design.

Formality compares the points between two designs and tries to match them using different algorithms including name-based matching and non-name-based matching algorithms. Based on Formality's user guide [4], it first compares the points based on their exact names. Then, it tries to perform case-insensitive name mapping or filtering out some characters. Name matching can also be done through mapping driven/driving nets (name of nets) of points. In the second phase, it attempts to match the remaining unmatched points using topological analysis of the unmatched cones. In other words, it matches two points with different names if they have equivalent structures. The final step is signature analysis which is based on generating functional and topological signatures. Functional signatures use random patterns simulation to generate primary outputs' data or register's output data to match different points. However, if an adversary inserts a hard-to-detect hardware Trojan, signature analysis may incorrectly match points since their simulation result is same. As a result, Formality may not be able to detect inserted Trojans (as indicated in Table 5.1). The proposed method is based on polynomial manipulation of different regions of the circuit and it is not dependent on the simulation or pattern generation. Thus, the proposed method outperforms Formality when there are hard-to-activate Trojan in the implementation.

Table 5.1 Trojan localization using Trust-HUB benchmarks

Benchmark			FANCI [19]	Formality [16]	This approach				False positive%			Improvement	
Type	#Gates	#TrojanGates	#SuspGates	#SuspGates	#SuspGates	#Spolys	#Ipolys	CPU time(s)	Proposed	[19]	[16]	[19]	[16]
RS232-T1000	311	13	37	214	13	62	186	0.67	0	24	201	a	a
RS232-T1100	310	12	36	213	14	61	189	0.86	2	24	201	12×	100.5×
S15850-T100	2456	27	76	710	27	592	1888	1.13	0	49	683	a	a
S38417-T200	5823	15	73	2653	26	1667	5004	3.12	11	58	2638	5.27×	239.8×
S35932-T200	5445	16	70	138	22	1778	4441	3.18	6	54	122	9×	20.33×
S38584-T200	7580	9	85	47	11	840	3905	4.74	2	76	38	38×	19×
Vga-lcd-T100	70,162	5	706	b	22	2426	7572	38.97	17	701	b	41.23×	b

[a]The proposed approach does not produce any false positive gates (infinite improvement)
[b]Cases that Formality could not detect the Trojans

5.3.2   Trojan Localization

Table 5.1 presents results for hardware Trojan localization. The first three columns show the type of benchmarks, number of gates in the circuit, and number of malicious gates (consisting of Trojan trigger and payload), respectively. The fourth column shows the number of suspicious gates reported by "FANCI" [19] approach. FANCI reports 1% to 8% of circuit nodes as false positive nodes on average (we have reported suspicious nodes as false positive nodes plus actual Trojan gates). The fifth column shows the number of suspected gates that can be found using Formality. It reports some faulty flip-flops or primary outputs which may have different values because of change in the functionality. However, there are so many gates in the cone corresponding to the faulty primary outputs or flip-flops and all of these gates are suspicious. In case Vga-lcd-T100, the Trojan effects are masked due to observability issues and nature of the above-mentioned signature analysis, and Formality returns no suspicious nodes. The sixth column shows the number of suspicious gates that the proposed method finds. This method detects all of the Trojan circuit gates (no false negative gates) plus very small number of false positive nodes (benign gates). The seventh column shows the number of specification polynomials which is equal to the number of flip-flops in the design plus the number of primary outputs. The eighth column presents the number of implementation polynomials which is equal to the number of fanout-free cones existing in the implementation. The CPU time (in seconds) to localize the Trojan is reported for each benchmark in ninth column. The time complexity of this method is linear with respect to the number of gates. The tenth, eleventh, and twelve columns show the number of false positive gates that the proposed approach, FACNI [19], and Formality [16] report, respectively. Clearly, this approach returns only few false positive gates. We are aware of the fact that comparison with FANCI is not fair since it does not require golden model. However, FANCI returns a lot of suspicious gates that it may not include all of the Trojan gates. For example, FANCI has reported top twenty suspicious gates for S35932-T200, none of them are from Trojan gates. Moreover, FANCI returns a set of suspicious gates even when the circuit is Trojan free. The next columns show the improvement in comparison with FANCI and Formality based on the number of false positive gates. This approach has a significant improvement compared to existing approaches—this approach reports orders of magnitude less false positive gates compared to [19] and [16].

5.3.3   Test Generation

For test generation, we used TetraMAX [17], the ATPG tool from Synopsys to generate tests exhaustively to activate the reported suspicious nodes. Since the suspicious candidates are few, we can exhaustively check several combinations to activate the Trojan. However, without using the localization method or using heuristic methods such as [19], exhaustive method will not work due to large number of suspicious gates. Table 5.2 shows the number of tests needed for activation and

Table 5.2 The required tests to activate the Trojan

Benchmark	N = 1			N = 2			N = 4		
	W/O local-ization	With local-ization	Improvement	W/O local-ization	With local-ization	Improvement	W/O local-ization	With local-ization	Improvement
RS232-T1000	311	13	23.9×	48,205	78	618.0×	4E+8	715	5E+5×
RS232-T1100	310	14	22.1×	47,895	91	526.3×	4E+8	1001	4E+5×
S15850-T100	2456	27	91.0×	3E+6	351	8.6E+3×	2E+12	17,550	9E+7×
S38417-T200	5823	26	224.0×	2E+7	325	5.2E+4×	5E+13	14,950	3E+9×
S35932-T200	5445	22	247.5×	1E+7	231	6.4E+4×	4E+13	7315	5E+9×
S38584-T100	7580	11	689.1×	3E+7	55	5.2E+5×	1E+14	330	4E+11×
Vga-lcd-T100	70,162	22	3189.2×	2E+9	231	1.1E+7×	1E+18	7315	1E+14×
Average	13,155.28	19.85	640.97×	2.9E+08	194.57	1.6E+6×	1.4E+17	7025.14	1.4E+13×

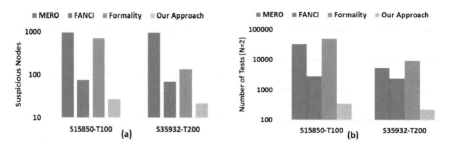

Fig. 5.4 (**a**) Number of suspicious nodes, (**b**) Number of tests needed to activate Trojans

detection of Trojans with/without using the localization method. First column shows the type of benchmark. The next two columns present the number of required tests to activate trigger conditions one at a time without and with using the localization method, respectively. The next column shows the improvement compared to without using localization. The proposed approach improves the number of required test vectors significantly. The next columns show the number of required tests to activate trigger conditions of two and four nodes at a time without and with using the localization method and the associated improvements, respectively. As it can be seen from Table 5.2, it is impractical to generate tests to activate four-node triggers even for these small benchmarks without the localization approach. If the localization is utilized, the number of required tests is reasonable and would be less by several orders of magnitude.

We also compared with MERO [3] for benchmarks S15850-T100 and S95932-T200. We did not compare using the remaining benchmarks because [3] did not report data for those benchmarks. Figure 5.4a shows the number of suspicious gates reported by the proposed approach compared to MERO. Clearly, this approach provides up to 44 times (40 times on average) reduction in suspicious gates compared to MERO. Figure 5.4b compares the number of tests required to activate the Trojan. As shown in the figure, this approach requires up to two orders of magnitude (60 times on average) less test vectors compared to MERO.

The experimental results demonstrate four important aspects of this approach. First, the number of false positive gates is very small and in some cases there are no false positives. In these cases, this method is able to detect the whole Trojan circuit. Next, all of the Trojan payload and trigger gates are inside the list of suspicious gates. In other words, this approach does not produce any false negative result. This approach detects both sequential and combinational Trojan circuits. Finally, this approach generates very few suspicious nodes (less than 0.2% of original design, less than 0.03% in most cases) that enable us to exhaustively generate tests to activate various trigger conditions to detect the Trojan circuit.

5.4 Summary

In this chapter, we presented an automated approach to localize functional Trojans in third-party IPs. First, we identified whether a third-party IP contains malicious functionality or it is trustworthy. Next, we presented an algorithm to localize the suspicious area of the Trojan-inserted IP to a region which contains very few (less than 0.03% of the original design in most cases) gates. This approach does not require any unrolling or simulation of the design and it formally identifies the parts of the circuit that is Trojan free as well as the remaining suspicious gates. In order to further aid in Trojan detection, we proposed a greedy test generation method to activate the Trojan. The experimental results demonstrated the effectiveness of the proposed methodology on Trust-HUB benchmarks. The localization approach reduces the overall Trojan detection effort (number of tests) by several orders of magnitude compared to the existing state-of-the art techniques.

References

1. J. Aarestad, D. Acharyya, R. Rad, J. Plusquellic, Detecting Trojans through leakage current analysis using multiple supply pad I_{ddq}s, in *IEEE Transactions on Information Forensics and Security* (IEEE, New York, 2010), pp. 893–904
2. B. Çakir, S. Malik, Hardware Trojan detection for gate-level ICS using signal correlation based clustering, in *Proceedings of the 2015 Design, Automation & Test in Europe Conference & Exhibition* (EDA Consortium, San Jose, 2015), pp. 471–476
3. R.S. Chakraborty, F. Wolf, C. Papachristou, S. Bhunia, MERO: a statistical approach for hardware Trojan detection, in *International Workshop on Cryptographic Hardware and Embedded Systems (CHES'09)* (Springer, Berlin, 2009), pp. 369–410
4. *Formality, User Guide* (2007). http://www.vlsiip.com/formality/ug.pdf
5. X. Guo, R.G. Dutta, Y. Jin, F. Farahmandi, P. Mishra, Pre-silicon security verification and validation: a formal perspective, in *ACM/IEEE Design Automation Conference (DAC)* (ACM, New York, 2015)
6. M. Hicks, M. Finnicum, S. King, M. Martin, J. Smith, Overcoming an untrusted computing base: Detecting and removing malicious hardware automatically, in *IEEE Symposium on Security and Privacy (SP)* (IEEE Computer Society, Los Alamitos, 2010), pp. 159–172
7. Y. Huang, S. Bhunia, P. Mishra, MERS: Statistical test generation for side-channel analysis based Trojan detection, in *Proceedings of the 2016 ACM SIGSAC Conference on Computer and Communications Security* (ACM, New York, 2016), pp. 130–141
8. Y. Huang, S. Bhunia, P. Mishra, Scalable test generation for Trojan detection using side channel analysis. IEEE Trans. Inf. Forensics Secur. **13**(11), 2746–2760 (2018)
9. Y. Jin, Y. Makris, Hardware Trojan detection using path delay fingerprint, in *Hardware-Oriented Security and Trust (HOST)* (IEEE, Piscataway, 2008), pp. 51–57
10. Y. Lyu, P. Mishra, A survey of side channel attacks on caches and countermeasures. Springer J. Hardw. Syst. Secur. (HASS) **2**(1), 33–50 (2018)
11. Y. Lyu, P. Mishra, Efficient test generation for Trojan detection using side channel analysis, in *Design automation and test in Europe (DATE)* (IEEE, Piscataway, 2019)
12. S. Narasimhan, X. Wang, D. Du, R. Chakraborty, S. Bhunia, TeSR: a robust temporal self-referencing approach for hardware Trojan detection, in *Hardware-Oriented Security and Trust (HOST)* (IEEE, Piscataway, 2011), pp. 71–74

13. M. Oya, Y. Shi, M. Yanagisawa, N. Togawa, A score-based classification method for identifying hardware-trojans at gate-level netlists, in *Design Automation and Test in Europe(DATE)* (Association for Computing Machinery, New York, 2015), pp. 465–470
14. A. Sayed-Ahmed, D. Gro, M. Soeken, R. Drechsler, et al., Formal verification of integer multipliers by combining gröbner basis with logic reduction, in *2016 Design, Automation & Test in Europe Conference & Exhibition (DATE)* (IEEE, Piscataway, 2016), pp. 1048–1053
15. C. Sturton, M. Hicks, D. Wagner, S. King, Defeating UCI: Building stealthy and malicious hardware, in *IEEE Symposium on Security and Privacy (SP)* (IEEE Computer Society, Los Alamitos, 2011), pp. 64–77
16. *Synopsys, Formality* (2015). http://www.synopsys.com/Tools/Verification/FormalEquivalence/Pages/Formality.aspx
17. *Synopsys, Tetramax ATPG*. http://www.synopsys.com/Tools/Implementation/RTLSynthesis/Test/Pages/TetraMAXATPG.aspx
18. *Trust-HUB*. https://www.trust-hub.org/
19. A. Waksman, M. Suozzo, S. Sethumadhavan, FANCI: identification of stealthy malicious logic using Boolean functional analysis, in *ACM SIGSAC Conference on Computer & Communications Security* (ACM, New York, 2013), pp. 697–708

Chapter 6
Vulnerability Assessment of Controller Designs

6.1 Introduction

Ensuring the integrity of an IC is challenging due to the diversity of attacks and attack goals. Malicious modifications [11], side-channel attacks such as power analysis [13] and timing analysis [12], debug infrastructure vulnerabilities [1], and fault injection attacks [2] can be exploited to affect security of a hardware design. A design can be resilient against such vulnerabilities when the security is considered from early design stages including controller and datapath design efforts.

Wide variety of solutions are proposed to protect datapath components [6, 8, 10, 19]. However, only a few studies addressed potential integrity issues of control circuits. Control circuits are required to be resilient against different types of attacks since they are responsible for controlling the functionality of the overall design and any deviation from the expected behavior can lead to severe impacts on security of the whole design. A finite state machine (FSM) of a secure design usually contains protected states which control proper handling of secret information. Fault injection attacks [2], existing EDA tools incompleteness [5] as well as designers' mistakes can compromise the security of a control circuit. An attacker's goal is to utilize existing FSM vulnerabilities to bypass authorized states and access the protected states illegally to weaken the security of the design or leak secret information such as cryptographic keys. Sunar et al. have shown that the secret key of RSA encryption algorithm [3] can be leaked when fault injection attack is used against the implementation of the Montgomery ladder algorithm [21]. It has been shown that some FSM encodings are more vulnerable toward fault injection attacks and an adversary can use the existing encoding vulnerabilities to have unauthorized access to the protected states [16]. Therefore, it is vital to identify and remove the vulnerabilities in the FSM architecture to protect them against any susceptibilities.

There are limited efforts to identify and address the security vulnerabilities of a control circuit. Sunar et al. used triple module redundancy (TMR) and parity checking methods to protect FSM of encryption algorithms against fault injection

© Springer Nature Switzerland AG 2020
F. Farahmandi et al., *System-on-Chip Security*,
https://doi.org/10.1007/978-3-030-30596-3_6

attacks [21]. However, the proposed technique introduces large area overhead (200%) and cannot detect other adversarial models such as hardware Trojans and vulnerabilities introduced by synthesis tools. In [22], a multilinear code selection algorithm is used to make cryptographic algorithm robust against fault injection attacks. However, this technique is not resilient against fault injection vulnerabilities caused by synthesis tools [5]. It has been shown that synthesis tools may insert additional "don't care" states in implementation of FSMs by using RTL don't care conditions and create assignments to optimize the gate-level netlist. At the same time, an adversary can use don't care states as a backdoor to access protected states and weaken the security of the overall design. In [5], authors use reachability as a trust metric to identify gate-level paths to protected states which do not exist in the RTL design. However, authors do not evaluate actual vulnerabilities caused by don't care states. They proposed an architectural change to state flip-flops in order to remove the access to the protected states from unprotected ones. Their proposed solution limits the functionality of the design. In [7], authors used mutation testing to detect existing hardware Trojans in unspecified functionality. However, mutation testing is very slow, and it may require significant manual intervention. Nahiyan et al. have proposed a state reachability analysis using ATPG tools [16]. They generate test patterns using the principle of n-detect-test [14] to extract the state transition graph (STG) of a given circuit. However, this option does provide any guarantees, e.g., in case one of their benchmarks they could not extract the whole STG. Sun et al. have proposed an FSM traversal technique using symbolic algebra [20]. However, their technique can only check the reachable states from a given state (e.g., initial state) and their technique cannot detect don't care states that may be introduced by synthesis tools. Similarly, they cannot detect hardware Trojans inserted in FSMs outputs.

In this chapter, we present a scalable formal approach that enables efficient FSM anomaly detection in state transition functions as well as FSM outputs. The proposed method models the specification of a given FSM as a set of polynomials ($\mathbb{F}_{spec}$) such that each polynomial is responsible for describing all of the valid states that can be reached. Each output of the FSM also can be represented using one specification polynomial. The specification polynomials can be derived from RTL codes as well as design documents. We also partition the gate-level implementation of an FSM based on the boundary of flip-flops, primary inputs, primary outputs, and fanout-free regions. We model each region by a polynomial and add it to the set of implementation polynomials ($\mathbb{F}_{imp}$). In the next step, we use Gröbner basis theory [4] to check the equivalence between two sets $\mathbb{F}_{spec}$ and $\mathbb{F}_{imp}$. We reduce each specification polynomial F_{spec_i} using a set of implementation polynomials. If the reduction leads to a non-zero remainder, there are some vulnerabilities in implementation of F_{spec_i}. Every assignment that makes the remainder non-zero reveals the conditions that can activate the hidden malfunction.

This approach is fully automated and it is guaranteed to find hard-to-detect FSM vulnerabilities in the implementation of an FSM when existing equivalence checking approaches fail. Experimental results demonstrate the effectiveness of the approach. Figure 6.1 shows the overall flow for anomaly detection using equivalence

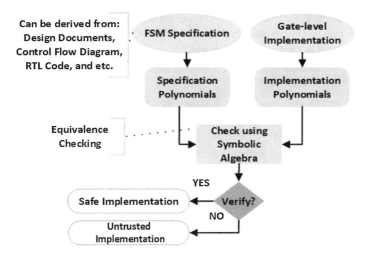

Fig. 6.1 Overview of FSM anomaly detection approach

checking. We demonstrated the merit of this proposed method by detecting the vulnerabilities in various FSM designs, while state-of-the-approaches failed to identify the security flaws.

The rest of the chapter is organized as follows. Section 6.2 illustrates how the proposed approach detects FSM vulnerabilities. We show the effectiveness of the approach using the experimental results in Sect. 6.4. Finally, conclusion is provided in Sect. 6.5.

6.2 Finite State Machine Anomaly Detection

A state machine can be defined with six characteristics: an initial state S_{init}, set of possible states $\mathbb{S}$ where $S_{init} \in \mathbb{S}$, set of possible input events $\mathbb{I}$, a state transition function (F_T) that maps combination of states and inputs to states ($F_T : \mathbb{S} \times \mathbb{I} \to \mathbb{S}$), a set of output events ($\mathbb{O}$), and an output function (F_O) that maps states and inputs to outputs ($F_O : \mathbb{S} \times \mathbb{I} \to \mathbb{O}$). Based on the function F_T which defines transitions, each state S_i can be accessed through a set of immediate, authorized states as well as a set of specific input events. Set $\mathbb{A}_{S_i} = \{(S_j, I_j)|S_j \in \mathbb{S} \ \& \ I_j \in \mathbb{I}\}$ shows legal conditions to access state S_i and set $\mathbb{A}_{\mathbb{S}}$ shows all of the legal ways to access states $\mathbb{S}$. If state S_i can be accessed through the condition (S_m, I_m) where $(S_m, I_m) \notin \mathbb{A}_{S_i}$, it is a threat to the integrity of the design. In other words, state S_i should not be accessed through some illegal conditions/states which do not exist in the specification. From the security perspective, it is important that a design exactly performs as intended in the specification, nothing more nothing less. The extra access path to state S_i, (S_m, I_m) may endanger the integrity of the design as it may create a backdoor to access the critical secrets/assets. The extra access paths will create extra states and transitions to the gate-level implementation of the FSM which do not exist in the behavioral specification. Formally, $\mathbb{A}_\mathbb{S}$ can be converted to $\mathbb{A}'_\mathbb{S}$. The extra set of access paths to the states of $\mathbb{S}$ can be computed as: $\mathbb{A}_M = \mathbb{A}'_\mathbb{S} - \mathbb{A}_\mathbb{S}$.

Example 6.1 The state transition diagram of a simple FSM is shown in Fig. 6.2. The FSM has three states: G, C, and protected state O representing with binary encoding 01, 10, and 00 respectively as shown in Fig. 6.2. The FSM is responsible for checking a password before starting a specific operation. Operation state (O) should be accessed only from check password state (C) when a password is entered, and it is valid ($a = 1$ and $b = 1$). An adversary may use the unspecified conditions to insert illegal transitions to gain access to the operation state (protected state) from the state G without even entering the correct password to bypass the security protection ($a = 1$ and $b = 0$). On the other hand, the synthesis tool or the designer mistake can also introduce some unintentional illegal access ways (don't care states D) to the protected state and compromise the security of the design. With respect to the specification, $\mathbb{A}_O$ should be equal to: {$(C, "a = 1 and b = 1")$. However, there are illegal access ways to state O in FSM implementation which is equal to: $\mathbb{A}_{M_O} = \{(D, "a"), (G, "a = 1 and b = 0")\}$. An adversary can compromise the security of the design by exploiting the existing vulnerabilities and attack the FSM. One of the possible attacks is fault injection attack [16]. The strategy is that the attacker tampers operating characteristics such as clock signal frequency, operating voltage, or working temperature hoping to change different path delays and force the FSM to capture next state incorrectly. One example would be to force the FSM to go to the don't care states which have access to protected states or attack target states. For instance, an attacker can inject a fault during transition 01 → 10 ($G → C$) to end up in don't care state 11 which has an immediate access to the protected state O and bypass password checking process in Example 6.1. The other possible attack is that the adversary inserts hardware Trojan by manipulating state transition graph in order to access certain states when a specific input event is triggered. In this case, the adversary is considered as an in-house rogue designer or an untrusted vendor/foundry. For instance, Example 6.1 shows that an adversary has inserted a Trojan that provides an illegal access way to state O from state G. The Trojan is typically hard-to-activate (from the unspecified design space) with negligible effect on the design constraints such as area and power to avoid detection from existing verification and debug flow.   ∎

Based on above observations, any deviation of FSM implementation from the specification (including extra access ways) can endanger the overall design integrity. In the rest of this chapter, we propose a promising approach to analyze FSMs to find potential malicious functionality. In this chapter, we consider illegal access paths as threat model, and the goal is to identify them using symbolic algebra.

Although the presented approach of Chap. 4 is promising for verification of arithmetic circuit, applying it on a general sequential circuit is challenging due to several reasons. First, formulating the specification of a general circuit cannot be

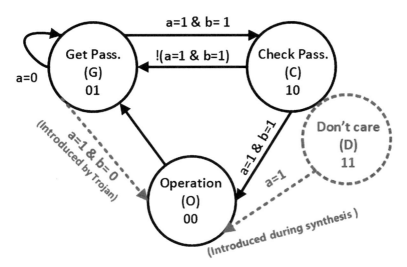

Fig. 6.2 The state diagram for checking a password in order to perform a specific operation. Potential vulnerabilities are shown with dotted lines

modeled as one simple and comprehensive polynomial. The specification may be modeled as a set of polynomials. However, finding the corresponding parts which are only responsible for implementing a special specification polynomial is not straightforward. Second, the implementation of a sequential circuit is not acyclic and it contains several loops which make the reduction operation infinite. Finally, time unrolling of the implementation is not efficient since it increases the design complexity and makes the equivalence checking inefficient. Moreover, existing Trojan may be activated after a large number of cycles (since the trigger condition is rare); therefore, there is no specific information about the required number of unrolling. In this chapter, we try to address the above-mentioned challenges to apply symbolic algebra to verify the trustworthiness of any general FSM. We not only check the given FSM for the correct expected behavior, but we also analyze the FSM to find any potential malicious extra access ways that may endanger the security of the FSM (nothing more). Finding extra access path especially from don't care states cannot be found using any formal methods such as model checkers since they are not accessed through the normal operation path. The remainder of this section describes the different parts of the approach: deriving specification polynomials, generating implementation polynomials, and performing equivalence checking in order to ensure the correctness of implementation and finding potential extra vulnerabilities.

6.2.1 Deriving Specification Polynomials

The specification of an FSM can be extracted from its state transition diagram or from a high-level description of the design (e.g., HDL modules). State transition graph can be derived from the design documentation as well as other high-level behavioral description of FSM such as RTL codes. In other words, deriving specification polynomials does not require a golden design/netlist.

Modeling the whole FSM using only one specification polynomial is not possible without considering the time notation in the specification polynomial as transitions between different states may be dependent on binary values of a specific input variable over different clock cycles. For example, as it is shown in Fig. 6.2, state C can be accessed from path $G \to C$ when in two consecutive clock cycles t_1 and $t_1 + 1$ such that $a = 0$ in t_1 and $a = 1, b = 1$ in $t_1 + 1$. Writing these conditions as a polynomial (part of the overall specification polynomial) without considering the timing will lead to a zero polynomial as $(1 - a).a.b = 0$. However, if we add timing notations to variables, the implementation also has to be time unrolled to match with the specification which increases the complexity of the equivalence checking problem. As a result, representing the functionality of an FSM using one specification polynomial is not possible. We propose an approach to model the specification of the FSM using polynomials without time unrolling the design.

Transitions of an FSM can be decomposed as: $F_T = \bigcup_{i=1}^{n} \mathbb{A}_{S_i}$, where n is the number of states and $\mathbb{A}_{S_i}$ shows all of the possible access ways of state S_i and F_T is the transition function of the FSM. To derive a set of specification polynomials which represent the whole FSM, we model each of $\mathbb{A}_{S_i}$ as one polynomial representing the legal access ways to state S_i and we add it to the set $\mathbb{F}_{spec}$.

A valid transition to state S_i happens when the current state is one of the authorized states and the corresponding input conditions are valid. In other words, S_i will be reached in the next clock cycle when the current state is S_j and condition $C_{j \to i}$ where $(S_j, C_{j \to i}) \in \mathbb{A}_{S_i}$ are evaluated to true. Note that, we show the value of variable x in the next cycle using x' notation. Therefore, transition $S_j \to S_i$ is modeled to a polynomial as: $f_{S_j \to S_i} : S_i' - (S_j.C_{j \to i}) = 0$. The polynomials of each of the conditions in $\mathbb{A}_{S_i}$ should be XORed to derive a polynomial representing the whole $\mathbb{A}_{S_i}$ since only one of them should be valid at the same time. We illustrate the approach using Example 6.2.

Example 6.2 In order to extract specification polynomials for FSM shown in Fig. 6.2, we consider each of the states independently and write a polynomial to represent conditions which update the next value of the state. For example, state O should only be accessed from state C when $a = 1$ and $b = 1$ or when the current state is state O and input a is equal to one. Since it should be accessed only from one of these conditions at a time, the conditions should be XORed to each other to show the effect of one condition at a time (the only exception is the condition of $a = 0$ in state G that will be ORed to other conditions since it works as the reset signal). The

O' shows the next value of state O. The specification of the FSM shown in Fig. 6.2 can be modeled as a set of three abstract polynomials ($\mathbb{F}_{spec} = \{f_G, f_C$ and $f_O\}$) as shown in Eq. 6.1.   ■

$$\mathbb{F}_{spec} : \{f_G : G' - ((1 - a) \vee (C.(1 - a.b) \oplus O)) =$$
$$G' - (1 - a + a.O + 2.a.b.C.O + a.C - 2.a.C.O - 1.a.b.C) = 0$$
$$f_C : C' - a.b.G = 0$$
$$f_O : O' - (a.b.C) = 0\}$$

(6.1)

We will describe how specification polynomials are used to check security properties of an FSM in Sect. 6.2.3. Before performing the equivalence checking, we need to refine specification polynomials to apply proposed FSM equivalence checking process since the proposed method requires that specification variables' names be the same as the corresponding variables in the implementation. We refine specification polynomials based on the FSM encoding style as well as corresponding names of state flip-flops in the implementation (name mapping between flip-flop names and corresponding variables in specification polynomials). We refine the variables which represent states in specification polynomials based on naming and encoding information that can be found in the high-level description of the design such as RTL modules as we describe in Example 6.3. As a result, the specification of FSM outputs can also be modeled with word-level specification polynomials based on state variables as well as primary inputs.

Example 6.3 Suppose that the RTL code shown in Listing 6.1 is the RTL version of the state machine shown in Fig. 6.2. We can see that states G, C and O are encoded as $\{01, 10, 00\}$ respectively. The state variable and next states are presented using variables $\{s_0, s_1\}$ and $\{n_0, n_1\}$. Therefore, the variables shown in Eq. 6.1 can be updated based on the above-mentioned information. For instance, variable G and next state variable G' can be modeled as $(1 - n_1).n_0$ and $(1 - s_1).s_0$, respectively. As a result, the specification polynomials shown in Eq. 6.1 can be rewritten as shown in Eq. 6.2. Note that, considering C encoded as $s_1.(1 - s_0)$ and O as $(1 - s_1).(1 - s_0)$, the terms $-2.C.O$ as well as $2.a.b.O.C$ of F_G in Eq. 6.1 are evaluated in updated specification polynomials).   ■

$$\mathbb{F}_{spec} : \{f_G : (1 - n_1).n_0 - (1 - a.b.s_1 + a.b.s_0.s_1 - a.s_0) = 0$$
$$f_C : n_1.(1 - n_0) - (a.b.(1 - s_1).s_0) = 0,$$
$$f_O : (1 - n_1).(1 - n_0) - (a.b.s_1.(1 - s_0)) = 0\}$$

(6.2)

Specification polynomials can be extracted directly from the RTL modules by using some specific rules. The logical operations in "If" statements can be mapped to polynomials. For example, by considering the encoding, line $G : if(a == 1`b1\&\&b == 1`b1)n <= C$ can be modeled as equation $n_1.(1 - n_0) = a.b.(1 - s_1).s_0$ In the next step, the corresponding polynomials of "If Then Else" are XORed together to achieve the exclusive nature of these statements. The derived specification polynomials will be used in the equivalence checking procedure.

Listing 6.1 RTL module of FSM shown in Fig. 6.2

```
module fsm(input clock, a, b; output valid );
reg[1:0] s, n;
parameter O=2'b00, G=2'01, C=2'b10;
always @(a, b, s) begin
case(s)
  G: if(a == 1'b1 && b == 1'b1) begin
    n <= C;
  end else if (a == 0) begin
        n <= G; end
  C: if(a == 1'b1 && b=1'b1 )
        n <= O;
  else
    n <= G; end
  O: n <= G;
end
always @(posedge clock)
begin
  if(a==1'b0) s <= G;
  else s <= n; end
end
endmodule
```

6.2.2   Generation of Implementation Polynomials

The goal is to partition the design and find the regions that are responsible for implementing each of the states and represent them as implementation polynomials. In order to perform this task, a mapping between state names and their corresponding gate-level state flip-flop names is needed. Here, we assume that the name of state inputs, outputs as well as state flip-flops are same between specification (RTL, state diagram, etc.) and implementation, or name mapping can be done based on existing methods in [15]. For the ease of the illustration, we explain how to extract the implementation polynomials when the FSM encoding is binary encoding. The proposed approach works for any state encoding.

After name mapping, we partition the gate-level implementation of the FSM based on state flip-flops. The state region construction starts from the input of the corresponding state flip-flop. The region construction continues with the inputs of the state flip-flop and moves backward recursively until it reaches to primary inputs or flip-flop outputs. The constructed region is converted to a polynomial by converting each of its gates to a polynomial as shown in Eq. 4.1 and combining them to each other to create one polynomial representing the whole region. We illustrate the approach using Example 6.4.

Example 6.4 Figure 6.3 shows the gate-level netlist which implements the FSM shown in Fig. 6.2. In the implementation, FSM states are encoded using binary scheme (two flip-flops are used to implement the functionality of three states shown

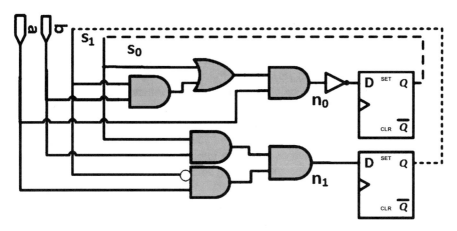

Fig. 6.3 Implementation of FSM in Fig. 6.2 using binary encoding

in the state diagrams of Fig. 6.2). The implementation is partitioned starting from the input of state flip-flop n_i and it is continued until reaching either primary inputs or outputs of state flip-flops (s_i). In the next step, the corresponding polynomial of each partition is derived by combining polynomials of each gate in the region to represent the functionality of next state variables (n_i). The implementation polynomials are shown in Eq. 6.3.   ∎

$$\mathbb{F}_{imp} : \{n_0 - (1 - a.b.s_1 - a.s_0 + a.b.s_0.s_1) = 0,$$
$$n_1 - (a.b.s_0 - a.b.s_0.s_1) = 0\}$$

(6.3)

When a gate's output goes to more than one gate, it is called a fanout. A fanout-free region is a set of gates that are directly connected together. Therefore, we partition the implementation to fanout-free regions and model each of them as one polynomial. The corresponding polynomials of each next state variable (n_is) can be computed by combining the polynomials of the corresponding fanout-free regions. Polynomials of fanout-free regions are calculated in order to reduce the efforts of implementation polynomial generation since one fanout-free region may be used in constructing the functionality of several n_is. Note that, in the implementation shown in Fig. 6.3, the functionality of each n_i is constructed with only one fanout-free cone.

Note that, the implemented functionality of FSM's outputs also can be formulated as a function of FSM inputs and states and presented as polynomials. In order to find implementation polynomials corresponding to FSM's outputs, each output gate is considered and traversed backward until it reaches to either input/output of state flip-flops or FSM inputs. The traversed gates are modeled using one polynomial showing the functionality of the corresponding output, and those polynomials are added to set $\mathbb{F}_{imp}$.

6.2.3 *Equivalence Checking*

From the security point of view, it is important to make sure that the implementation of a design performs exactly its specification. We check the functional equivalence between a control logic specification and its implementation in order to establish the trust of the control logic. In this chapter, we formulate the FSM equivalence checking as ideal membership testing based on Gröbner Basis theory. Implementation polynomials $\mathbb{F}_{imp}$ are formed as an ideal I based on particular order $>$ (the topological order which exists in the implementation). FSM implementation is trustworthy if all of the specification polynomials in set $\mathbb{F}_{spec}$ are the member of ideal $I =< \mathbb{F}_{imp} >$.

In order to check the trustworthiness of the implementation, each specification polynomial F_{spec_i} from set $\mathbb{F}_{spec}$ is reduced over polynomials in $\mathbb{F}_{imp}$. All of the variables in specification polynomials (except primary inputs and flip-flops' outputs) are substituted with the corresponding functionality of the variable from the implementation polynomials. Note that, the reduction procedure is done using sequential polynomial division as shown in Sect. 4.2.2. The reduction process continues until a zero remainder or a non-zero polynomial which contains a combination of primary inputs and flip-flop outputs is reached. If reduction F_{spec_i} over set $\mathbb{F}_{imp}$ results in a zero remainder, it means that F_{spec_i} belongs to the ideal $I =< \mathbb{F}_{imp} >$. In other words, set $\mathbb{F}_{imp}$ has successfully implemented the specification F_{spec_i}. Otherwise, the implementation of F_{spec_i} is not trustworthy (implementation is not equal to specification). If all of the remainders are equal to zero polynomials, it means that the overall implementation is equal to FSM's specification since set $\mathbb{F}_{spec}$ includes specification of the FSM states as well as specification of FSM's outputs (specification polynomials cover all specification space). Algorithm 13 shows the equivalence checking procedure.

Algorithm 13: FSM equivalence checking algorithm

Input: Gate-level netlist imp and specification polynomials $\mathbb{F}_{spec}$
Output: FSM anomalies $\mathbb{E}$
$\mathbb{F}_{imp}$=findImplementationPolynomials(imp)
for *each* $f_{spec_i} \in \mathbb{F}_{spec}$ **do**
    r_i = reduction of f_{spec_i} over $F_j s \in \mathbb{F}_{imp}$
    if *($r_i != 0$)* **then**
        $\mathbb{T}_i = findNonZeroAssignments(r_i)$
        $\mathbb{E}.put(f_{spec_i}, \mathbb{T}_i)$
    end
end
return $\mathbb{E}$

Algorithm 13 takes the gate-level netlist imp of a given FSM as well as the specification polynomials $\mathbb{F}_{spec}$ as inputs and tries to find any existing anomalies in the FSM. First, it computes the implementation polynomials ($\mathbb{F}_{imp}$) as described in Sect. 6.2.2 (line 4). In the next step, every specification polynomial f_{spec_i} (corresponding to state S_i) in $\mathbb{F}_{spec}$ is reduced over a set of implementation polynomials F_js using Gröbner Basis theory in order to find the remainder r_i (line 6). If the remainder is non-zero, it means that there are some malicious functionality in implementing specification polynomial f_{spec_i}. Every assignments that make the remainder non-zero activates the malicious access path to S_i. The algorithm stores the anomalies in the map $\mathbb{E}$ (lines 7–9).

Example 6.5 Consider the specification polynomials of Eq. 6.2, gate-level netlist in Fig. 6.3 as well as implementation polynomials shown in Eq. 6.3. Equation 6.4 shows the equivalence checking procedure with respect to topological order $\{n_1, n_0\} > \{s_1, s_0, a, b\}$. Note that, reducing of variables $\{n_1, n_0\}$ happen at the same time as their orders are the same. However, we show the reduction of F_{spec_1} in two steps to illustrate the procedure better. ∎

$$F_{spec_1} : f_G : (1 - n_1).n_0 - (1 - a.b.s_1 + a.b.s_0.s_1 - a.s_0)$$
$$stp_{11} : (1 - a.b.s_0 + a.b.s_0.s_1).n_0 - (1 - a.b.s_1 + a.b.s_0.s_1 - a.s_0)$$
$$stp_{12} : (1 - a.s_0 - a.b.s_1 + a.b.s_0.s_1) - (1 - a.b.s_1 + a.b.s_0.s_1 - a.s_0) = 0$$
$$F_{spec_2} : f_C : n_1.(1 - n_0) - (a.b.(1 - s_1).s_0)$$
$$stp_{21} : (-a.b.s_0.s_1 + a.b.s_0) - (a.b.(1 - s_1).s_0) = 0$$
$$F_{spec_3} : f_O : (1 - n_1).(1 - n_0) - (a.b.s_1.(1 - s_0))$$
$$stp_{31} : (a.s_0 - a.b.s_0 + a.b.s_1) - (a.b.s_1 - a.b.s_1.s_0) =$$
$$(remainder) : a.s_0 - a.b.s_0 + a.b.s_0.s_1$$

$$(6.4)$$

As shown in Eq. 6.4, specification polynomials of states G and C are reduced to zero which means that they are safely implemented by the gate-level netlist. However, the reduction of specification polynomial of the protected state O results in a non-zero remainder. The remainder reveals potential vulnerabilities in the gate-level implementation of the design to access the protected state O. Every assignment that makes the remainder non-zero discloses an unauthorized access path to the state O. Table 6.1 shows the malicious access paths. As it can be observed from Table 6.1, don't care state $\{s_1, s_0\} = 2`b11$ can access the protected state O due to synthesis tool optimization (when input a is true). There is another malicious access path to the state O from state G when $a = 1$ and $b = 0$. This extra access is a hardware Trojan that was inserted by an adversary or a rogue designer.

Table 6.1 Malicious access paths to the protected state O shown in Fig. 6.2

s_1	s_0	a	b
1	1	1	X
0	1	1	0

6.2.4 FSM Security Property Checking

There may be some signals in the implementation such as *reset* signal that indirectly influences the state transition but considering it in specification polynomials make them very complex. In this section, we are proposing a security property checking technique to formally verify the security properties of a given FSM. In other words, we want to make sure that there is no unintentional mistakes or hidden FSM Trojans (hidden among other functionalities of the gate-level netlist) that endanger the integrity of the overall implementation. In the proposed approach, we perform efficient property equivalence checking in order to verify security properties of a given state. We consider security properties such that there should be no additional access path to a given state other than the access paths that are listed in the specification. In other words, we want to formally verify that a given state S_j (e.g., a protected state) cannot be accessed when all of the valid transitions $\mathbb{A}_{S_i}$ are not active. To formally verify that, we generate the specification polynomials as described in Sect. 6.2.1 for a given FSM. For each specification polynomials f_{spec_i}, we compute every assignment to the variables that make the whole specification polynomial evaluated to zero. We expect that in the presence of such assignments (conditions), the corresponding implementation polynomial is also reduced to a zero polynomial. However, if applying such assignment leads to a non-zero remainder, there is a malicious access path to the given state S_i. Algorithm 14 presents the proposed approach.

Algorithm 14 takes gate-level netlist *imp* of a given FSM with state space $\mathbb{S}$ as well as the specification polynomials $\mathbb{F}_{spec}$ as inputs and finds the anomalies exist in the FSM. First, it finds the corresponding specification of polynomial f_{spec_i} for each state $S_i \in \mathbb{S}$ (line 5). In the next step, the algorithm tries to find all of the assignments that make f_{spec_i} zero and put them in set $\mathbb{N}$ (line 6). For each assignment N_j, we extract the corresponding implementation polynomial F_j representing the behavior of implementation of S_i under conditions N_j (line 8). We expect F_j is reduced to a zero remainder for a safe implementation. However, if F_j is reduced to a non-zero polynomial, there is a malicious access path to S_i when all of the valid transactions to S_i are inactive. Every assignment that makes the F_j (remainder) non-zero activates the malicious access path to S_i. The algorithm stores the anomalies in the map $\mathbb{E}$ (lines 9–15). In the proposed approach, instead of performing the common equivalence checking that only check the implementation of $\mathbb{A}_{S_i}$ for state S_i, we check that there is no other transition except the transition listed in $\mathbb{A}_{S_i}$ to access state S_i (nothing more). Since we check everything else except the valid transitions, the checking may seem exponential. However, we control the size of the problem by assigning don't care values to existing variables in the given specification polynomials. Moreover, by assigning values, the complexity of constructing implementation polynomials is decreased. Example 6.6 illustrates the approach.

Example 6.6 Consider the protected state O shown in Fig. 6.2 as well as the corresponding specification polynomial (property) shown in Eq. 6.2 which is equal to $(1 - n_1).(1 - n_0) = (a.b.s_1.(1 - s_0))$. We want to make sure that there is no malicious access path to protected state O, inserted either by synthesis tool or a

---

Algorithm 14: FSM security property checking algorithm

Input: Gate-level netlist imp, state $\mathbb{S}$ and specification polynomials $\mathbb{F}_{spec}$
Output: FSM anomalies $\mathbb{E}$
for *each $S_i \in \mathbb{S}$* **do**
  $f_{spec_i} = \mathbb{F}_{spec}.get(S_i)$
  $\mathbb{N}$=findZeroAssignments(f_{spec_i})
  for *each $N_j \in \mathbb{N}$* **do**
    F_j=findImplementationPolynomial(imp, S_i, N_j)
    if *($F_j! = 0$)* **then**
      $T_j = findNonZeroAssignments(F_j, N_j)$
      if *($\mathbb{E}.get(f_{spec_i})$) == null)* **then**
        $\mathbb{E}.put(f_{spec_i}, T_j)$ **else**
      end
      $\mathbb{T}_i = \mathbb{E}.get(f_{spec_i})$
      $\mathbb{E}.put(f_{spec_i}, (T_j \cup \mathbb{T}_i)$
    end
  end
 end
end
return $\mathbb{E}$

---

rogue designer. In the first step, we find the assignments that make the right side of the equation $(1 - n_1).(1 - n_0) = (a.b.s_1.(1 - s_0))$ zero. The assignments are listed in Table 6.2. In the next step, we construct an implementation polynomial modeling the implementation of state O for each of the assignments listed in Table 6.2 using the gate-level netlist shown in Fig. 6.3. The property checking procedure is shown in Eq. 6.5. ∎

$$\{s_1, s_0, a, b\} = \{0XXX\} : f_{O1} = (1 - n_1).(1 - n_0) = a.s_0 - a.b.s_0$$
$$\{s_1, s_0, a, b\} = \{11XX\} : f_{O2} = (1 - n_1).(1 - n_0) = a$$
$$\{s_1, s_0, a, b\} = \{100X\} : f_{O3} = 0$$
$$\{s_1, s_0, a, b\} = \{1010\} : f_{O4} = (1 - n_1).(1 - n_0) = 0$$

(6.5)

As it can be seen in Eq. 6.5, condition $\{s_1, s_0, a, b\} = \{0XXX\}$ generates a non-zero polynomial f_{O1}. The assignment that makes f_{O1} non-zero (we call it partial remainder) is $\{s_1, s_0, a, b\} = \{0110\}$ which shows the malicious extra path from state G to state O as shown in Fig. 6.2. Moreover, condition $\{s_1, s_0, a, b\} = \{11XX\}$

Table 6.2 Conditions that
inactivate all of the transitions
to state O shown in Fig. 6.2

s_1	s_0	a	b
0	X	X	X
1	1	X	X
1	0	0	X
1	0	1	0

generates a non-zero polynomial f_{O2}. The assignment that makes f_{O2} non-zero is
$\{s_1, s_0, a, b\} = \{111X\}$ which shows the malicious extra path from don't care state
$\{s_1, s_0\} = 11$ caused by synthesis tool to state O as shown in Fig. 6.2. Using the
values to construct implementation polynomials not only enables us to perform
security property checking on FSM designs but also controls the size of partial
remainders and makes the overall approach scalable.

6.3   Effect of Encoding on FSM Vulnerabilities

As we mentioned earlier, most security vulnerabilities in an FSM are unintentionally
created by designer mistakes or by CAD tools. Traditional FSM design practices
are driven by cost and performance while security is largely ignored. For example,
FSMs are generally encoded in binary, gray or one-hot from the performance perspective. In [16], it was shown that certain encoding schemes are more susceptible to
fault injection attacks. Further, CAD tools can create additional vulnerabilities in an
FSM. In this section, we describe how vulnerabilities are introduced by traditional
FSM encoding schemes.

Binary Encoding In binary encoding scheme, states are encoded as a binary
sequence where the states are numbered starting from 0 and up. The number of
state flip-flops (FF_S), q, required for binary encoding scheme is given by $q = log_2(n)$; where, n is the number of states. From this equation, it is evident that
binary encoding scheme requires minimum number of state FFs. Therefore, binary
encoding scheme is better suited for FSM with a fewer number of states. However,
in terms of security, the binary encoding scheme makes the FSM more susceptible
to fault injection attack since any fault can create a valid state transition.

One-Hot Encoding In one-hot encoding, only one bit of the state variable is "1"
while all other state bits are zero. One-hot encoding requires as many state FFs as
the number of states, and therefore, one-hot encoding requires more state FFs than
binary. From the security perspective, it is inherently less vulnerable to fault attacks
since the probability of injecting one fault and ending up in a valid state is low (two
faults are needed). On the other hand, one-hot encoding could result in many don't
care states. If any of these don't care states has access to a protected state as a result
of synthesis, then there will be a vulnerability in the FSM.

Gray Encoding In gray encoding, consecutive states only differ by one bit. Gray code may require the same number of state FFs and combinational logic just as complex (if not more) as binary encoding. From the security perspective, gray encoding makes the FSM susceptible to fault attacks (similar to binary encoding). In this section, we present two security-aware FSM encoding techniques. The first approach (Scheme I) is based on making protected states more resilient against fault attacks using combined benefits of one-hot and binary encodings, while the second approach is used to secure prohibited transitions instead of every transition.

In this section, we present two security-aware FSM encoding techniques. The first approach (Scheme I) is based on making protected states more resilient against fault attacks using combined benefits of one-hot and binary encodings, while the second approach is used to secure prohibited transitions instead of every transition.

Scheme I One-hot encoding is more resilient to fault injection attacks in comparison with other encoding styles as discussed earlier. The first encoding scheme exploits the benefits of one-hot style while reducing the number of don't care states. Algorithm 15 shows the proposed encoding. The algorithm takes as input from the designer, the states specified as three different categories: the initial state, normal states, and protected states. The primary goal is to make the protected states more resilient against fault attacks. Therefore, the algorithm uses one-hot scheme for protected states while it uses binary scheme for normal states. If the FSM contains one initial state, N normal states, and P protected states, the algorithm uses $log(N) + P$ bits for encoding (line 5). The algorithm dedicates P upper bits to one-hot scheme while it pads zero for the rest of $log(N)$ bits in order to encode a protected state (lines 7–9). To encode a normal state, the algorithm pads zero for the N upper bits and uses binary encoding for $log(N)$ lower bits (lines 10–12). It always encodes the initial state with all zeros (line 13). This encoding approach decreases the number of don't care states (as compared to one-hot) while making sure that it will be impossible for an attacker to access to a protected state from a normal state with fault attacks since during normal state transitions, P upper bits are fixed to zeros.

Example 6.7 The FSM in the controller circuit of SHA-256 digest engine is shown in Fig. 6.4b. The FSM is composed of 7 states: "Reset," "Data Input," "Padding," "Block Process," "Block Next," "Valid," and "Error." Each of these states controls specific operations in the SHA-256 digest engine. The digest algorithm operates on two registers, $w[0..64]$ which is responsible for loading the message and $h[0..7]$ which stores the intermediate digest results. These two registers are initialized during "Reset" state. The final digest (H) will be latched into the result register in "Valid" state. In the SHA-256 FSM example, "Valid" is a protected state and "Block Next" is the authorized state to access the protected state "Valid." The FSM shown in Fig. 6.4b can be securely encoded with Algorithm 15 as following: Reset="00000," Block Process="00001," Block Next="00010," Padding="00011," Error="00100," Data Input="01000," Valid="10000."

Algorithm 15: Secure encoding—scheme I

Input: Protected States $\mathbb{P}$, Normal States $\mathbb{N}$, Initial State I
Output: FSM Encoding Map, S_{EN}
$P \leftarrow |\mathbb{P}|, N \leftarrow |\mathbb{N}|$
$l = P + log(N)$ Encoding Bit Length
for $P_i \in \mathbb{P}$ **do**
$\quad |\quad E_i = OneHotEncoding(i, P)||(00..0)_{l-p}$ concat (l-p) zeros with one-hot encoding
$\quad |\quad S_{EN}.add(P_i, E_i)$
end
for $N_i \in \mathbb{N}$ **do**
$\quad |\quad E_i = binaryEncoding(i, N)||(00..0)_{l-N)}$
$\quad |\quad S_{EN}.add(N_i, E_i)$
end
$S_{EN}.add(I, (00..0)_l)$ initial state
return S_{EN}

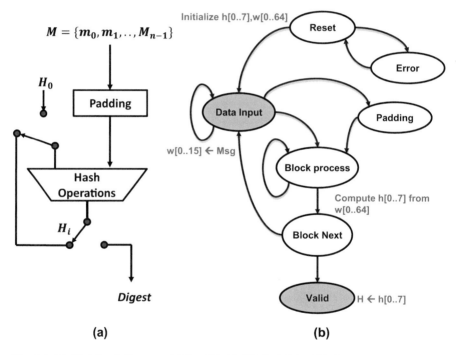

Fig. 6.4 (a) High-level diagram of SHA-256, (b) FSM of SHA-256 digest engine. Red States show the protected states in FSM of SHA-256 digest [17]

Scheme II Note that, every access to a protected state from an unauthorized state does not necessarily introduce a security threat based on the attack objective. For example, it can be observed from Fig. 6.4b that an unauthorized access to *Data Input* state from *Block Process* may not be a security threat if the attacker's objective is to bypass the digest operation. In other words, an FSM may be secured against fault attacks if the state encoding provides protection for only prohibited transitions instead of every transition to the protected states. This property enables us to introduce another FSM encoding scheme which is similar to binary scheme, but also tries to reduce the number of don't care states that exist in the previously proposed encoding (Algorithm 15).

Algorithm 16 shows the second proposed encoding approach. The algorithm takes an initial state, state names, and a list of the prohibited transitions as inputs and generates an optimal length encoding as output. A list of prohibited transitions includes state(s) that should be prohibited during a transition from state u to v using fault attacks. Moreover, it can contain information about which transitions should not be bypassed. If there are n states, the algorithm searches different encoding lengths (l) where $log(n) \le l \le n - 1$ and tries almost all of the combinations to find a secure encoding (lines 4–8). The goal is to find an encoding that does not have any conflict with the list of prohibited transitions. The initial state is encoded with all zeros. To check whether an attacker can inject a fault during a transition from state u to v and gain access to state t, a mask is generated from the temporary encodings of states u and v to identify which bits have changed during this transition (line 10). The changed bits are marked with "x" and the fixed bits are kept as they are in the generated mask (e.g., "0101"→"1001": $mask =$"xx01"). The encoding of state t is compared with the generated mask. If the encoding has one-bit difference from the fixed bits of the mask, the temporary assignment is safe (since reaching to t requires changes in the fixed bits of transition $u \rightarrow v$). Otherwise, the assigned encodings are not safe and another combination should be tried (lines 11–12). The algorithm returns an encoding as a result when there is no conflict with the list of prohibited transitions (lines 13–14). Note that we also employ some heuristics to efficiently reduce the computation cost of the algorithm (e.g., using one-hot scheme in l bits and assign it to l states to limit the search space). If there is an optimal encoding, this algorithm will find it. In the worst case, it uses one-hot scheme for all of the states except the initial state like the previous approach. However, this approach requires more inputs from the designer.

Example 6.8 Using Algorithm 16, the FSM shown in Fig. 6.4 can be securely encoded as: Reset="0000," Block Process="1000," Block Next="0100," Padding="0010," Error="0111," Data Input="0001," Valid="1110." Both of these encoding showed in this example and Example 6.7 protect critical states of SHA-256 algorithm (shown in red in Fig. 6.4) toward fault injection attacks.

Algorithm 16: Secure encoding—scheme II

Input: State Names $\mathbb{S}$, Initial State I, Prohibited Transitions $\mathbb{T}$
Output: FSM Encoding Map, S_{EN}
for $log(N) \leq l \leq N - 1$ **do**
 for *all of possible combinations* **do**
  $S_{EN} = \{\}$
  $S_{EN}.add(I, (00..0)_l)$ initial state
  $S_{EN} = findEncoding(\mathbb{S}, l)$ random encoding with length l
  for $T_i \in \mathbb{T}$ **do**
   m=generateMask($S_{EN}.get(src(T_i))$, $S_{EN}.get(dest(T_i))$)
   for *prohibited states t_i of T_i* **do**
    | checkForConflicts($S_{EN}.get(t_i)$, m);
   end
  end
  if *(There is no conflict)* **then**
   | **return** S_{EN}
  end
 end
end

6.4   Experiments

6.4.1   Experimental Setup

In order to evaluate the effectiveness of the FSM anomaly detection approach, we have implemented the proposed algorithms using Java. The experiments were run on a PC with Intel core i7 and 16 GB memory. We have applied this method on various FSM benchmarks from "OpenCores" [18]. The benchmarks are described using RTL modules (that we treat as the specification). To obtain the gate-level implementation, we synthesize RTL modules using "Synopsys Design Compiler" [9]. We extract specification polynomials from RTL modules of FSM benchmarks considering their state transitions and output assignments. We have implemented a Java program such that we define the valid transitions to states in the form of abstracted polynomials and it generates one specification polynomial representing all of the logical transitions to a given state. The same approach was used to produce the specification polynomials for FSM outputs. On the other hand, implementation polynomials are driven automatically from the synthesized gate-level netlist using the proposed framework. In order to generate implementation polynomials, gate-level netlist is partitioned into the fanout-free regions which are restricted to flip-flops boundaries as well as primary input and primary outputs. We use fanout-free regions to reduce the number of implementation polynomials. We reduce specification polynomials over a set of implementation polynomials and each non-zero remainder represents an FSM security threat. The goal is to find the assignments to activate the vulnerabilities (if any).

Table 6.3 Result of the proposed FSM anomaly detection technique using equivalence checking

Benchmark	Encoding	#Gates	#FF	#States	#Trans.	DC Sts	DC Tran.	EQ (s)
TAP controller	One-hot	136	16	16	33	3	6	80.63
AES encryption	One-hot	88	5	5	11	0	0	6.26
AES encryption	Binary	60	3	5	11	3	6	5.03
RSA encryption	One-hot	114	7	7	9	0	0	18.48
RSA encryption	Binary	76	3	7	9	1	1	6.2
SHA digest	One-hot	153	7	7	47	121	121	50.89
Multiplier controller	Binary	52	3	5	8	3	3	1.85
SAP controller	Binary	135	4	12	25	0	0	17.23

6.4.2 Results

We have conducted two sets of experiments based on whether the vulnerability is introduced by the synthesis tool (unintentional) or an attacker (intentional). In the first set of experiments, the gate-level implementations are Trojan-free, and all the potential vulnerabilities are caused by the synthesis tool. Note that different encoding styles and values can create different vulnerabilities. In the second set of experiments, we have inserted hardware Trojans in state transitions as well as state outputs of the implementations in order to show the effectiveness of this approach. The results are shown in Table 6.3 and Fig. 6.5, respectively.

Table 6.3 represents the result of proposed FSM equivalence checking approach for eight different benchmarks. The first column shows the type of the benchmark. The second column represents the encoding style of the FSM design. We have considered binary and one-hot encoding methods to show that the proposed approach is not dependent on the encoding approach. The third, fourth, and fifth columns represent the number of gates, number of state flip-flops, and the number of states, respectively. The sixth column represents the number of transitions in the FSM design. The next two columns indicate the number of don't care states and don't care transitions that the method finds, respectively. Note that this method does not report the don't care states that are not connected to any other states. Finally, the last column shows the CPU time that the proposed equivalence checking (EQ) approach to find anomalies in FSM benchmarks.

To show that the proposed approach can also detect hardware Trojans inserted in the state transition function as well as in the logic that generates the outputs of the FSM, we inserted hardware Trojans by exploiting the unspecified functionality of different benchmarks. Figure 6.5 shows the required time to detect the injected Trojan. The attributes of the benchmarks are the same as shown in Table 6.3.

The experimental results demonstrated that the approach could detect the hidden vulnerabilities introduced by synthesis tool optimization while Formality fails to detect them. Note that some state encodings are more likely to have vulnerabilities caused by synthesis tools. For example, the synthesis tools tend to map all of the

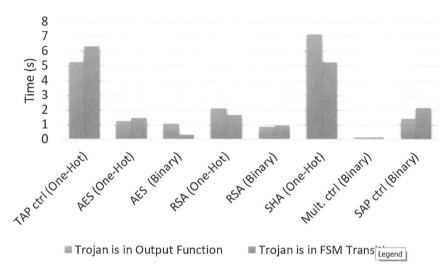

Fig. 6.5 Time required to detect hardware Trojans in output logic and state transition function

don't care states to a state with all zero's encoding (e.g., 3'b000) assuming that the state represents reset or ideal state. If the protected state is mapped using this encoding, there may be a direct access to the protected state from some don't care state caused by the synthesis tool.

6.5   Summary

It is critical to make sure that FSMs are correctly implemented, and there is no deviation from the specified functionality of the FSM since any unexpected functionality can endanger the integrity of the whole design. FSM vulnerabilities can be caused intentionally through an adversary by inserting hardware Trojan in the implementation or unintentionally using CAD tools such as synthesis tools. In this chapter, we presented an approach to formally detect anomalies in finite state machines using symbolic algebra. The proposed approach models the specification of an FSM as a set of polynomials such that each polynomial represents all of the valid transitions to one of the states of the FSM. We modeled the implementation of an FSM as a set of polynomials. We check the equivalence of the specification polynomials and implementation polynomials using Gröbner basis theory. We have showed the approach can detect hidden vulnerabilities created by both synthesis tools or an adversary.

References

1. J. Backer, D. Hély, R. Karri, Secure design-for-debug for systems-on-chip, in *IEEE International Test Conference (ITC)* (IEEE, Piscataway, 2015), pp. 1–8
2. E. Biham, A. Shamir, Differential fault analysis of secret key cryptosystems, in *Annual International Cryptology Conference* (Springer, Berlin, 1997), pp. 513–525
3. E. Brickell, A survey of hardware implementations of RSA, in *Advances in Cryptology-CRYPTO89 Proceedings* (Springer, Berlin, 1990), pp. 368–370
4. D. Cox, J. Little, D. O'shea, in *Ideals, Varieties, and Algorithms*, vol. 3 (Springer, Berlin, 1992)
5. C. Dunbar, G. Qu, Designing trusted embedded systems from finite state machines. ACM Trans. Embed. Comput. Syst. (TECS) **13**(5s), 153, 2014
6. F. Farahmandi, Y. Huang, P. Mishra, Trojan localization using symbolic algebra, in *Design Automation Conference (ASP-DAC), 2017 22nd Asia and South Pacific* (IEEE, Piscataway, 2017), pp. 591–597
7. N. Fern, K.-T.T. Cheng, Detecting hardware trojans in unspecified functionality using mutation testing, in *Proceedings of the IEEE/ACM International Conference on Computer-Aided Design* (IEEE, Piscataway, 2015), pp. 560–566
8. X. Guo, R.G. Dutta, P. Mishra, Y. Jin, Scalable SoC trust verification using integrated theorem proving and model checking, in *IEEE International Symposium on Hardware Oriented Security and Trust (HOST)* (IEEE, Piscataway, 2016)
9. https://www.synopsys.com/support/training/rtl-synthesis/design-compiler.html
10. Y. Huang, S. Bhunia, P. Mishra, MERS: statistical test generation for side-channel analysis based trojan detection, in *Proceedings of the 2016 ACM SIGSAC Conference on Computer and Communications Security* (ACM, New York, 2016), pp. 130–141
11. R. Karri, J. Rajendran, K. Roseland, M. Tehranipoor, Trustworthy hardware: identifying and classifying hardware trojans, in *IEEE Computer* (IEEE, Piscataway, 2010), pp. 39–46
12. P.C. Kocher, Timing attacks on implementations of diffie-hellman, RSA, DSS, and other systems, in *Annual International Cryptology Conference* (Springer, Berlin, 1996), pp. 104–113
13. P. Kocher, J. Jaffe, B. Jun, Differential power analysis, in *Annual International Cryptology Conference* (Springer, Berlin, 1999), pp. 388–397
14. S.C. Ma, P. Franco, E.J. McCluskey, An experimental chip to evaluate test techniques experiment results, in *Proceedings, International Test Conference, 1995* (IEEE, Piscataway, 1995), pp. 663–672
15. T. Meade, S. Zhang, Y. Jin, Netlist reverse engineering for high-level functionality reconstruction, in *2016 21st Asia and South Pacific Design Automation Conference (ASP-DAC)* (IEEE, Piscataway, 2016), pp. 655–660
16. A. Nahiyan, K. Xiao, K. Yang, Y. Jin, D. Forte, M. Tehranipoor, AVFSM: a framework for identifying and mitigating vulnerabilities in FSMS, in *Design Automation Conference (DAC), 2016 53nd ACM/EDAC/IEEE* (IEEE, Piscataway, 2016), pp. 1–6
17. A. Nahiyan, F. Farahmandi, D. Forte, P. Mishra, M. Tehranipoor, Security-aware FSM Design Flow for Identifying and Mitigating Vulnerabilities to Fault Attacks, in *IEEE Transactions on Computer-Aided Design of Integrated Circuits and Systems (TCAD)* (IEEE, Piscataway, 2018)
18. *OpenCores*. http://opencores.org
19. J. Rajendran, V. Vedula, R. Karri, Detecting malicious modifications of data in third-party intellectual property cores, in *Proceedings of the 52nd Annual Design Automation Conference* (ACM, New York, 2015), p. 112
20. X. Sun, P. Kalla, F. Enescu, Word-level traversal of finite state machines using algebraic geometry, in *2016 IEEE International High Level Design Validation and Test Workshop (HLDVT)* (IEEE, Piscataway, 2016), pp. 142–149
21. B. Sunar, G. Gaubatz, E. Savas, Sequential circuit design for embedded cryptographic applications resilient to adversarial faults. IEEE Trans. Comput. **57**(1), 126–138 (2008)
22. Z. Wang, M. Karpovsky, Robust fsms for cryptographic devices resilient to strong fault injection attacks, in *2010 IEEE 16th International On-Line Testing Symposium* (IEEE, Piscataway, 2010), pp. 240–245

Chapter 7
SoC Security Verification Using Property Checking

7.1 Introduction

Functional properties have been widely used in pre-silicon verification to formally model the expected behavior of the design and its specification. Functional properties can significantly reduce verification and debug efforts as they can be used in formal verification tools (e.g., model checking tools) to automatically prove/disprove whether all aspect of design functionality have been correctly implemented and they are aligned with the specification. Functional properties can also be used in simulation-based validation to pinpoint the source of functional violations automatically [3]. Moreover, these properties can be synthesized and placed on silicon to monitor particular events at the run-time or provide closures for post-silicon validation [22]. Similar to functional properties, we need to create security properties for SoC security verification and validation to either prove the trustworthiness of the design or find a counterexample in the case of security violations in an automatic fashion. In other words, security properties help to provide provable guarantees against various vulnerabilities. Security properties formally describe the expected behaviors and rules that a trustworthy design is required to follow for each type of SOC vulnerability.

SoC security can be compromised through information leakage, timing and power side-channel attacks, implementation of malicious functionality [50], exploitation of design-for-test (DFT) and design-for-debug (DFD) infrastructures [15], fault injection attacks [38], and unsafe design transformations. These vulnerabilities can be introduced by untrusted third-party vendors, the rogue employees of the design house, an untrusted system integrator, or an unreliable foundry. These vulnerabilities can also be introduced unintentionally by designers' mistakes or lack of knowledge about the security requirements. Moreover, computer-aided design (CAD) tools can introduce additional security vulnerabilities in the design [18, 39]. These vulnerabilities can create a backdoor to leak sensitive information (e.g., encryption/decryption keys, random numbers, configuration bits,

© Springer Nature Switzerland AG 2020

F. Farahmandi et al., *System-on-Chip Security*,
https://doi.org/10.1007/978-3-030-30596-3_7

etc.) of the design, create a denial of service, or grant the control of the design to an attacker. Therefore, it is very important to identify the security vulnerabilities during design test and verification stages and address them as soon as possible while we still have the flexibility to change the design if needed. In this chapter, we show that how security properties can be extracted/generated for automatic security verification and validation of an SoC.

Once security properties are generated, they should be checked using various tools to ensure the secure behavior of the design. Security properties can be tested statically using formal tools such as model (property) checking [47], equivalence checking, [20], and information flow tracking techniques [25, 51] to formally provide proofs for lack of security vulnerabilities or identifying violations. Security properties can be checked dynamically as well by simulation. These properties can be synthesized and placed on silicon to monitor specific events at the run-time. Moreover, they can be mapped to a reconfigurable fabric to enforce security policies/rules during execution and ensure convenient upgradability to cover zero-day attacks. Additionally, security properties can be used for directed test generation to activate hard-to-detect security events [1–3, 6, 7, 11, 12, 16, 21, 28, 32, 33, 37, 42, 44]. The test generation technique will be described in Chap. 8.

The remainder of this chapter is organized as follows: Sect. 7.2 provides an overview on functional properties. Section 7.3 presents various properties that should be checked for vulnerability analysis of an SoC. Section 7.4 presents various tools to check security properties. Finally, Sect. 7.5 concludes the chapter.

7.2   Background: Writing Properties

A property in the context of verification is a statement that can check assumptions, conditions, and expected behaviors in a design. A property can be in the form of an assertion or cover statement. An assertion can check that if everything is working correctly in the design and notify if an illegal event has happened. Assertions can also be used for consistency checking [9]. A cover statement can check if a scenario has ever occurred in the design (during simulation or run-time execution). Therefore, cover statements can provide coverage information for design validation. There is a single bit associated with assert which indicates the pass or fail status of the assertion. The cover assertion triggers at the end of the execution when the assertion is not covered during the run-time. Properties can check design behaviors in two main ways: (1) immediate statements and (2) concurrent statements. An immediate property can check if the expected functional scenario is correct in a procedural block (similar to if-else statement) at an instance of time. In other words, an immediate assertion can check if a particular block of code has been executed if pass conditions are met. On the other hand, concurrent functional properties check for a design behavior over a period of time for the whole module (instead of a procedural block). Moreover, in concurrent properties, sampling of variables can occur in some instance of time and the evaluation of the whole property can be done during another time frame (called observe time). Concurrent assertions are more powerful and can be used to describe and check more complicated events.

These days, designers mostly use one of the powerful assertion languages such as PSL (property specification language) [23] and SystemVerilog Assertions [54] to describe interesting behavioral events of a design. These languages use temporal logic representations such as linear temporal logic (LTL) [40] and computational tree logic (CTL) [14]. Languages based on LTL and CTL usually describe design behaviors and properties in four layers: Boolean expression, sequence, property specification, and assertion directive layers. These layers can be used on top of different HDL languages including Verilog and VHDL. Boolean expressions are the most fundamental layer of a property which describe Boolean events on a signal or a combination of signals [52]. Logical operators AND, "&&", OR, "||", and NOT "!" can be used to evaluate Boolean expressions. Once Boolean expressions are evaluated, the sequence of these expressions is checked in the sequence layer. For example, property *(req) [wait for two cycles] (ack)* checks that if two Boolean expression $req == 1$ and $ack == 1$ are evaluated to true and $ack == 1$ happens two cycles after signal *req* is asserted.

Temporal sequences can be shown using different operators. Note that to show the operators, we use the notations used in SystemVerilog Assertions. However, the same operators can be found in other languages. Operator ## can be used to the number of clock cycles needed for an event to happen. For example, $a\#\#3b.ended$ shows that *b* completes three cycles after *a* happens. Operator ":" is used for concatenation and operators $[low : high]$, $[*]$, $[=]$, or $[->]$ present the notation of bounded or unbounded repetition. Operator $[*]$ shows the repetition of zero or more consecutive instances; however, $[=]$ and $[->]$ denote one or more non-consecutive repetitions. In $[=]$, the last value of the expression should not be necessarily true. For example, $a[= 2 : 3]$ means that *a* has been true for 2 or 3 non-consecutive clock cycles (with possible clock cycles at the end that *a* is not true). Sequences can be combined using several match operators such as "and," "intersect" (when both expressions are expected to be true at the same time), "or," "until," "throughout" (the first expression should be true at every clock cycle that the second expression is being evaluated), "whitin" (the first expression should be true at at least one time when the second expression is being evaluated), "s_eventually," etc.

Sequences can be combined to create another sequence or a property. Properties can be constructed using "not," "and," "or," "if... else," and implication "|− >" operators. For example, $a|-> b \quad until \quad (c||d)$ shows signal *a* is asserted, signal *b* must be asserted, and must stay true until one of the signals *c* or *d* is asserted. Properties can have different evaluation directives such as "assert" (making sure the described statement is always true), "cover" (checking whether the described scenario has happened during simulation), "assume" (assume this statement is true when evaluating other properties), and "expect" (making sure the described statement is always true in a procedural block). An active edge of the clock usually accompanies properties (usually the rising edge of the clock is considered as default). In general, properties can be classified into the three following groups, and different layers, operators, and directives help to construct them.

- Invariant properties describe conditions that should be always true (obligation mode) or never should be true (conditional mode). For example, A FIFO should not be written a new value when it is full.
- Sequential properties describe a set of conditions happening in a particular order over a period of time. For example, a "ready" signal should be followed in 2 or 3 clock cycles by the assertion of "enable" signal.
- Eventually properties describe a condition that should be followed by another condition in any number of clock cycles. For example, an access request to a shared bus should be eventually granted.

Example 7.1 Suppose that we have three design properties: first, whenever signal A asserted, signal B is supposed to be asserted within next three cycles. The following assertion describes this property:

```
property  A_1;
   @( posedge  clk )  (A  |->   [*1:3]  B);
endproperty
A1_assertion:  assert  property  (A_1);
```

Consider a second property where we would like to cover functional scenarios such that C and D signals are not true at the same time. This property can be formulated as:

```
property  A_2;
   @( posedge  clk )  (!  (C && D));
endproperty
A2_cover:  cover  property  (A_2);
```

The third property is signal E may be asserted only during the time frame beginning with signal F and continuing until signal K rises. This property can be formulated as:

```
property  A_3;
   E  [=0:$]  within  (F  ##[0:$]  $rose (K))
endproperty
A3_assertion:  assert  property  (A_3);
```

Note that $ denotes a finite but unbounded maximum. Function $rose() checks if signal K has changed from 0 to 1.   ■

For simplicity, we just show the core part of properties in the following sections of this chapter.

7.3 Creating Security Properties

Since designers may not have sufficient knowledge about the security requirements due to the huge complexity of SoC designs and their attack surfaces, it is difficult for them to manually analyze the design implementation in different levels of abstractions to identify potential vulnerabilities. Moreover, it is very costly for a design house to keep a large number of security experts that know all aspects of the design, and they can detect security issues. Therefore, it is necessary to generate a set of security properties that can be checked automatically to generate formal proof and check-marks for security assurance. To generate security properties, we convert each vulnerability to a set of rules and each rule to a set of properties. Here are some of the example rules:

- Asset confidentiality: the sensitive information of the design should not be leaked to observable points of the design, such as primary outputs, DFT, or DFD infrastructure.
- Memory and type safety: the content of memory should be protected from unauthorized modifications.
- Non-interference and isolation: the interaction of low-security entities with high-security entities should be protected to ensure that low-security entities are not able to observe any differences in the behavior of the system and high-security entities. Moreover, two entities with the same level of security should not be able to affect the integrity and confidentiality of each other.
- Resiliency toward side-channel attacks: timing dependencies of different components should be checked to prevent leakage of secret information through side-channel characteristics of the system. Hardware micro-architecture units such as branch predictors and speculation execution units should not leak the information of secret propagation in the design and create covert-channel attacks such as Spectre [29] and Meltdown [31].
- Resiliency toward fault injection attacks: different controller designs of an SoC should be resilient toward fault injection attacks to prevent changes in the flow of the design in order to skip some instructions or bypassing the security mechanisms of the design.

Mapping each vulnerability and corresponding rules to a set of security properties is a challenging task since a vulnerability may involve several SoC IPs and their interactions over multiple clock cycles. Generally, SoC security properties can be generated based on the following characterizations: (1) the type of vulnerability, (2) the type of the functionality and its granularity (IP-level, micro-architectural-level, and SoC-level), (3) design abstraction (RTL, gate-level, layout-level), and (4) time-to-check.

Based on the type of vulnerability, we may need to identify assets, critical data, and related information as well. In the next step, we select the IPs as well as SoC transactions that either contain/exhibit those vulnerabilities or involve in propagating the corresponding assets. For example, if we are concerned about information leakage, the focus of the generated properties should be on crypto IPs, TRNG modules, and asset management units. On the other hand, we need to focus on halt units and exception handlers in the SoC if we are concerned about denial of service attacks. In the next step, we create abstract properties that formulate the security rules for identified units to prevent vulnerabilities. Security properties formulation can be done in two general ways: (1) checking forbidden behaviors (i.e., conditional assertion statements) and (2) checking the correct implementation of expected security policies (i.e., assertions in obligation mode and cover statements) [22]. Next, we need to define the appropriate level of abstraction that those properties should be checked and tune the abstract security properties in a way to capture information at that level of abstraction. If a property targets a vulnerability that may happen in different levels of abstractions, we will use property mapping techniques to generate the same properties for other abstraction levels as well. Finally, we define the time that properties should be checked. Some properties should be statically checked at the design time using formal tools. For example, the registers which contain critical data should be accessed through only valid ways, and any undefined access to these registers is considered as a threat. Another example would be checking the initial values of intermediate buffers. Designers may decide to initialize some buffer values to don't cares for power and energy optimization purposes. However, if these buffers can control some critical functionality and be accessible from observable points of the design, their values can be maliciously set at the reset time by adversaries to bypass the security mechanisms of the system. Therefore, such properties can be statically checked using model checkers at the design time to find counterexamples of when they are violated. On the other hand, some properties should be checked at the run-time. For example, we need to create properties that check who can have access to the bus when encryption/decryption keys are transferred to corresponding units at the boot-time. We illustrate the security property generation approach with the following example.

Example 7.2 Suppose our design is the hardware implementation of AES encryption algorithm which has several components (e.g., the controller module, SBox, cipher and decipher blocks, key unit, etc.) as shown in Fig. 7.1. Our security goal is the absence of information leakage in the SoC. As a result, we need to focus on components that either perform security-critical applications or contain/propagate keys. The threat of information leakage can be considered as lack of confidentiality or integrity. As a result, two types of security properties can be generated:

1. Confidentiality:

    (a) Any linear function of key (K) value should not be leaked to the output (O)

$$no\_key\_leakage\_to\_output := assert(O \neq linear\_func(K))$$

    (b) Any linear function of key (K) value should not be leaked to design-for-test (DFT)

$$no\_key\_leakage\_to\_DFT := assert(DFT \neq linear\_func(K))$$

2. Integrity

    (a) Key register should be only accessed through valid ways:

$$Safe\_key\_register\_changeassert K_{access} \notin V = \{V_1, V_2, .., V_n\} | -> K_t == K_{t+1}$$

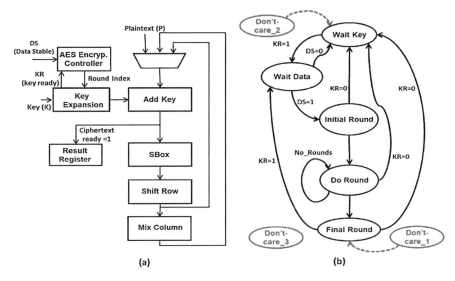

Fig. 7.1 AES encryption module. (**a**) AES encryption data path (**b**) FSM of the AES encryption controller [38]

The above assertion shows that the key register should be updated only through valid ways where V presents all the legitimate ways to update the key register (reset, from key management units, etc.). Otherwise, the key register should keep its value.

3. Side-channel vulnerability:

   (a) The secret key should not affect the cipher-text ready (R) signal otherwise there will be a side-channel leakage

   $$no\_side\_channel\_leakage := assert(P, K_t \neq K_{t+1})|- >$$

   $$(R == 0[*t : t + C + 1], R == 1, R == 0[*t + C : t + 2C - 1], R == 1)$$

   The above assertion shows that if we have a valid plain-text (P) and two valid encryption keys which will be received in cycles t and t+1, cipher-text ready signal should be asserted only in cycles t+C and t+2C. In other words, the different values of the key should not define when the cipher-text ready signal is asserted. The only parameter that defines the cipher-text ready signal is the number of rounds (C) in the implementation of the AES algorithm.

   Now, suppose that we need to check there are no hardware Trojans implemented in the AES implementation. Here is one sample assertion:

4. Malicious functionality:

   (a) There should not be any functional scenario where the plain-text (P) or any part of it is sent to cipher-text (O) unencrypted for all input assignments (I).

   $$no\_bypass\_encryption := assert(i \in I|- > (O \neq P))$$

Moreover, the following are some sample rules and properties that need to be checked regarding access controls in the controller design of the AES design.

5. Access controls:

   (a) The cipher-text ready signal (R) should be only asserted from Final Round state.

$$safe\_ready\_assertion := assert\$rise(R)|$$

$$- > \$past(state) == ``Final \quad Round"$$

   (b) Critical state Final Round should only be accessed through Do Round state.

$$Safe\_state\_access := assert(state == ``Final Round")|$$

$$- > \$past(state) = ``Do \quad Round"$$

The similar properties can be checked for the decryption unit.   ■

Each of the above-shown properties can be mapped to several properties when the abstract functions and symbols are replaced with real ones. In general, several properties can cover one security rule, and several rules can cover one security vulnerability. Security metrics can be used to not only identify vulnerabilities but also to define the security level of design. Metrics can also be used to establish security rules and properties. Moreover, the coverage of security properties can be used as a metric for determining the security level of the design. Figure 7.2 shows the relation between security vulnerabilities, rules, properties, and metrics.

Unique vulnerabilities can be introduced at different levels of design abstractions: RTL, gate-level, and layout. Therefore, some specific security rules and properties should be checked at those levels. For example, Property 1.b from Example 7.2 should be checked at gate-level since design-for-test and design-for-debug architecture can be defined at that stage. Similarly, properties 5.a and 5.b should be checked at RTL to ensure the correct implementation and lack of malicious modifications; they should also be tested at gate-level to make sure that the synthesis tool did not introduce any additional states during RTL to gate-level transformation [39]. Checking security rules and properties at different levels of abstraction may be the result of some other requirements such as scalability, flexibility, and the need for higher precision. For example, side-channel leakage properties can be checked at RTL if we need to address the potential vulnerabilities at the early stages of the design. However, if high precision is required, these rules and properties should be checked at gate-level or layout when the physical characteristics of the design are available.

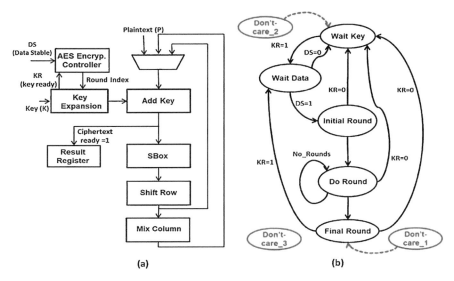

Fig. 7.2 The relation between security vulnerabilities, rules, properties, and metrics

7.4 Checking Security Properties

Different tools and techniques should be used to verify the extracted security properties. For example, integrity properties can be checked using model checking tools. On the other hand, confidentiality properties are easier to be checked using information flow tracking and taint analysis techniques since there are several ways that a secret value can be leaked to observable points of a design. Properties related to safe design transformation can be checked using equivalence checking tools.

7.4.1 Security Validation Using Model Checking

Model checking is a famous technique in design verification which checks a design for a set of given properties. To solve the model checking problem, the design and the given properties are converted to a mathematical model/language, and all of the design's states are checked to see whether the given properties are satisfied. A class of model checkers is designed based on temporal logic formula [14]. The properties are described using LTL formulas to specify the expected behaviors of the design. The properties are checked using the model checkers. A model checker either proves the correctness of a given property over all of the possible behaviors of the design or finds a counterexample when the property fails.

A model checker tries all of the possible states of a design to prove a given property using a binary decision diagram (BDD). However, the number of design states can be huge since every bit introduces two states in the design. For example, a

32-bit register can add 2^{32} states to the design state space. Although some techniques such as slicing, abstracting, etc. have been proposed [13, 53], state space explosion still is the primary limitation of using model checking in property verification. Bounded model checking (BMC) is introduced to address the state space explosion problem by selectively constructing and storing different states of a design for few cycles [5]. BMC tries to find a counterexample in the first K cycles during execution. If a counterexample is found within K cycles, the property does not hold. Otherwise, K can be increased in the hope of finding a counterexample in higher bounds. BMC is not able to prove a property since it unrolls the circuit for a specified number of clock cycles. However, it can provide a statistical metric for a given property when the model checker fails (e.g., no counterexample can be found in K clock cycles). The BMC problem can be mapped to satisfiability problem, and SAT-solvers can be utilized to solve the problem. Therefore, the BMC addresses some of the state space explosion problems associated with BDDs in model checking. Figures 7.3 and 7.4 show the model checking and bounded model checking approaches, respectively. Clearly, bounded model checking cannot provide proof for property P. However, it can reveal if the property P is violated within K clock cycles.

Security properties describe the expected behaviors which a trustworthy design is required to follow. Model checkers can be used to ensure safety properties. An SoC designer and a third-party vendor can agree on certain security properties that the design should satisfy. When the design is sent to the SoC integrator, the SoC integrator converts the design to a formal description to check the security properties using a model checker. If all of the security properties are verified, the expected security behaviors are met. Rajendran et al. have proposed a Trojan detection technique which is based on using bounded model checking [47]. They have considered the threat model as an attempt to corrupt the critical data such as secret keys of a cryptographic design, and random numbers which are required by most of the cryptography algorithms or stack pointer of a processor. The assumption is that these critical data should be stored in some specific registers and accesses to these registers should be protected. In other words, the registers which contain critical data should be accessed through valid ways, and any undefined access to

Fig. 7.3 Verification using model checking

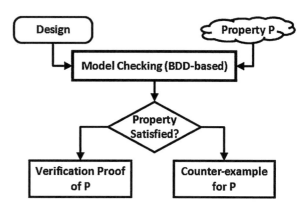

Fig. 7.4 Verification using bounded model checking

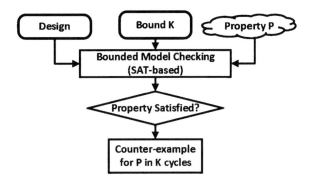

these registers is considered as a threat. The safe access conditions to these registers are formulated as properties (assertions), and a bounded model checker is utilized to find a counterexample when the security properties are violated.

Example 7.3 Suppose that the program counter (PC) register is considered as a critical data. The only valid ways to change the PC register is either using a reset signal (V_1), by CALL instruction which increments the PC register V_2 or using RET instruction which decrements the value of the PC register V_3. Otherwise, the PC register should keep its value. The safety property of PC register can be formulated as:

$$Safe\_PC\_change : assert \quad PC_{access} \notin \mathbb{V}=\{V_1, V_2, V_3\}|->\ \#\#1\,PC_t=PC_{t-1})$$

When this property is fed into a bounded model checker alongside with the processor design, a counterexample is expected to be found whenever PC register or a part of it is changed using an unauthorized access [47]. ∎

Security property checking can be done in two general ways: (1) checking forbidden behaviors and (2) checking expected security properties. The malicious behavior of design is formalized and checked using model checkers in [49]. The method can be applied only for known Trojan types. Hasan et al. have proposed a hardware Trojan detection technique using LTL and CTL security properties to generate hardware Trojan monitors in order to improve the resiliency of hardware designs against malicious functionality [24]. The attacker is considered as an untrustworthy third-party designer that can insert Trojans in the IP, and the defender is SoC integrator. The SoC integrator needs to formulate dangerous behaviors as security properties to perform vulnerability verification using model checkers. The generated counterexample and the involved signals are provided to the in-house designers to produce a guideline for efficient run-time security monitors.

The success of using model checking-based approaches to prove security properties is highly dependent on the size of the design, SAT-engine, and the quality of the given properties. The model checker cannot guarantee non-existence of security violations corresponding to vulnerabilites that we do not have properties to cover them. Coverage of security properties can provide a trust metric. Moreover, property checking can be used for automated generation of directed tests [8, 10, 17, 19, 27, 34–36, 41, 43, 45].

7.4.2   Security Property Checking using Information Flow Tracking

Researchers have proposed techniques based on formal methods to prove security-related properties that would violate confidentiality requirements. These methods are particularly effective for detecting information leakage inside cryptographic designs. One such method looks for confidentiality and integrity property violation [25, 26]. Confidentiality property requires that secret information never leaks to an unsecured domain, and integrity property requires that untrusted data never enters the secured domain. Information flow is traced by assigning a taint bit to it. In another approach [48], a base property is used to detect information leakage which may imply the existence of a Trojan. The base property checks whether any input sequence exists such that it triggers secret information leakage to an observable point. The security properties check whether there is an input assignment (or a sequence of input assignments) I which triggers the leakage of secret data S to output ports or observable points (O) of the design.

$$\exists i \in I \rightarrow (S == O)$$

The property and formal description of the design are fed into a bounded model checker to find the possible leakage. However, the above-mentioned property has several challenges. If the secret information S contains n bits, the model checker needs to check 2^n different values. Checking all possible values may not be feasible when n is in the order of hundred (which is normal for encryption algorithms). These kinds of properties can be checked using information flow tracking tools. A gate-level information flow tracking (GLIFT) [51] technique has been proposed to measure illegal flows of a tainted value at the Boolean level. This technique instruments the design with taint analysis logic for every gate in the design to track all implicit and explicit information flows. However, this approach introduces a huge design overhead as the size of the instrumented circuit grows quickly. Several methods have been proposed at the higher levels of abstractions to address this limitation [4]. Typed security languages have been also introduced to check information flow tracking using the structure of hardware description languages [30, 55]. However, the scalability of all of these techniques should be improved.

7.4.3   Equivalence Checking for Testing Security Properties

The security of a SoC can be compromised by exploiting the vulnerabilities of the finite state machines (FSMs) in the SoC controller modules through fault injection attacks. These vulnerabilities may be introduced by designer mistakes, Trojan insertion, or CAD tools. Potential threats introduced by third-party EDA tools are considered in [46]. It is possible that an adversary modifies a design using non-transparent EDA tools such as synthesis tools. A synthesis tool may optimize some registers and unsafely modify the FSM. The authors have proposed a hardware Trojan detection technique which is based on property coverage analysis to ensure that a gate-level netlist is free from hardware Trojans inserted by synthesis tools. The proposed Trojan detection method is based on both security property checking as well as state coverage to mark suspicious unused circuit states. Figure 7.5 shows the different ways to insert Trojans in an FSM.

Example 7.4 Consider the FSM shown in Fig. 7.5a. Whenever the current state is *A*, the next state should be either *A* or *B*. The property can be formulated using an LTL formula as shown below:

$$assert \quad always \quad (cur\_state==A)|->##1(next\_state=A|| \quad next\_state=B)$$

Note that, *X* symbol shows the next cycle and → shows implication.   ∎

The equivalence checking method presented in Chap. 6 can effectively check such properties since it formally identifies any additional access paths to FSM states (such as unauthorized access to a protected state), and generates counterexample in case of a security problem.

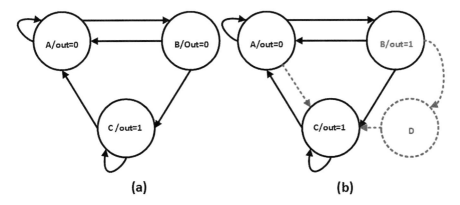

(a)   **(b)**

Fig. 7.5 Trojans in an FSM: (**a**) A Trojan-free FSM, (**b**) Trojan can be inserted into an FSM using different ways: (i) changing the state output (e.g., state *B*), (ii) modification of state transitions (e.g., extra transition from state *A* to *C*), and (iii) adding extra states (e.g., state *D*) and transitions (such as state transitions *B* → *D* and *D* → *C*) to FSM [38].

7.5 Chapter Summary

In this chapter, we presented several security properties to cover SoC vulnerabilities to perform automatic security verification. We discussed different tools should be used to check various properties effectively. Defining a comprehensive set of security properties allows detection of security violations and flows at earlier stages of design, which leads to cost minimization and easier adoption o required countermeasures.

References

1. A. Ahmed, P. Mishra, QUEBS: Qualifying event based search in Concolic testing for validation of RTL models, in *IEEE International Conference on Computer Design (ICCD)* (IEEE, Piscataway, 2017), pp. 185–192
2. A. Ahmed, F. Farahmandi, P. Mishra, Directed test generation using Concolic testing of RTL models, in *Design Automation and Test in Europe (DATE)* (IEEE, Piscataway, 2018), pp. 1538–1543
3. A. Ahmed, F. Farahmandi, Y. Iskander, P. Mishra, Scalable hardware Trojan activation by interleaving concrete simulation and symbolic execution, in *IEEE International Test Conference (ITC), Phoenix* (IEEE, Piscataway, 2018), pp. 1–10
4. A. Ardeshiricham, W. Hu, J. Marxen, R. Kastner, Register transfer level information flow tracking for provably secure hardware design, in *Design, Automation and Test in Europe Conference and Exhibition (DATE)* (IEEE, Piscataway, 2017), pp. 1695–1700
5. A. Biere, A. Cimatti, E.M. Clarke, O. Strichman, Y. Zhu, Bounded model checking. Adv. Comput. **58**, 117–148 (2003)
6. M. Chen, P. Mishra, Functional test generation using efficient property clustering and learning techniques. IEEE Trans. Comput. Aided Des. Integr. Circuits Syst. (TCAD) **29**(3), 396–404 (2010)
7. M. Chen, P. Mishra, Property learning techniques for efficient generation of directed tests. IEEE Trans. Comput. (TC) **60**(6), 852–864 (2011)
8. M. Chen, P. Mishra, Decision ordering based property decomposition for functional test generation, in *Design Automation and Test in Europe (DATE)* (IEEE, Piscataway, 2011), pp. 167–172
9. M. Chen, P. Mishra, Assertion-based functional consistency checking between TLM and RTL models, in *International Conference on VLSI Design* (IEEE, Piscataway, 2013), pp. 320–325
10. M. Chen, X. Qin, P. Mishra, Efficient decision ordering techniques for SAT-based test generation, in *Design Automation and Test in Europe (DATE)* (IEEE, Piscataway, 2010), pp. 490–495
11. M. Chen, P. Mishra, D. Kalita, Automatic RTL test generation from SystemC TLM specifications. ACM Trans. Embed. Comput. Syst. (TECS) **11**(2), article 38 (2012)
12. M. Chen, X. Qin, P. Mishra, Learning-oriented property decomposition for automated generation of directed tests. Springer J. Electron. Test. (JETTA) **30**(3), 287–306 (2014)
13. A. Cimatti, E. Clarke, F. Giunchiglia, M. Roveri, Nusmv: a new symbolic model checker. Int. J. Softw. Tools Technol. Transfer **2**(4), 410–425 (2000)
14. E.M. Clarke, E.A. Emerson, A.P. Sistla, Automatic verification of finite-state concurrent systems using temporal logic specifications. ACM Trans. Program. Lang. Syst. (TOPLAS) **8**(2), 244–263 (1986)
15. G.K. Contreras, A. Nahiyan, S. Bhunia, D. Forte, M. Tehranipoor, Security vulnerability analysis of design-for-test exploits for asset protection in SoCs, in *Asia and South Pacific Design Automation Conference (ASP-DAC)* (IEEE, Piscataway, 2017), pp. 617–622

16. J. Cruz, F. Farahmandi, A. Ahmed, P. Mishra, Hardware Trojan detection using ATPG and model checking, in *International Conference on VLSI Design (VLSI Design)* (IEEE, Piscataway, 2018), pp. 91–96

17. N. Dang, A. Roychoudhury, T. Mitra, P. Mishra, Generating test programs to cover pipeline interactions, in *ACM/IEEE Design Automation Conference (DAC)* (2009), pp. 142–147

18. C. Dunbar, G. Qu, Designing Trusted Embedded Systems from Finite State Machines. ACM Trans. Embed. Comput. Syst. (TECS) **13**(5s), 153:1–153:20 (2014)

19. F. Farahmandi, P. Mishra, Automated test generation for debugging arithmetic circuits, in *Design Automation and Test in Europe (DATE)* (IEEE, Piscataway, 2016), pp. 1351–1356

20. F. Farahmandi, P. Mishra, Automated debugging of arithmetic circuits using incremental gröbner basis reduction, in *2017 IEEE 35th International Conference on Computer Design (ICCD)* (IEEE, Piscataway, 2017), pp. 193–200

21. F. Farahmandi, P. Mishra, Automated test generation for debugging multiple bugs in arithmetic circuits. IEEE Trans. Comput. (TC) **68**(2), 182–197 (2019)

22. F. Farahmandi, R. Morad, A. Ziv, Z. Nevo. P. Mishra, Cost-effective analysis of post-silicon functional coverage events, in *Design, Automation and Test in Europe Conference and Exhibition (DATE)* (IEEE, Piscataway, 2017), pp. 392–397

23. M. Gruninger, C. Menzel, The process specification language (PSL) theory and applications. AI Mag. **24**(3), 63–74 (2003)

24. S.R. Hasan, C.A. Kamhoua, K.A. Kwiat, L. Njilla, Translating circuit behavior manifestations of hardware trojans using model checkers into run-time trojan detection monitors, in *IEEE Asian Hardware-Oriented Security and Trust (AsianHOST)* (IEEE, Piscataway, 2016), pp. 1–6

25. W. Hu, B. Mao, J. Oberg, R. Kastner, Detecting hardware trojans with gate-level information-flow tracking. Computer **49**(8) 44–52 (2016)

26. W. Hu, A. Ardeshiricham, M.S. Gobulukoglu, X. Wang, R. Kastner, Property specific information flow analysis for hardware security verification, in *2018 IEEE/ACM International Conference on Computer-Aided Design (ICCAD), San Diego, CA* (IEEE, Piscataway, 2018), pp. 1–8

27. H.-M. Koo, P. Mishra, Functional test generation using property decompositions for validation of pipelined processors, in *Design Automation and Test in Europe (DATE)* (IEEE, Piscataway, 2006), pp. 1240–1245

28. H.-M. Koo, P. Mishra, Functional test generation using design and property decomposition techniques. ACM Trans. Embed. Comput. Syst. (TECS) **8**(4), article 32 (2009)

29. C. Li, J. Gaudiot, Online detection of Spectre attacks using microarchitectural traces from performance counters, in *2018 30th International Symposium on Computer Architecture and High Performance Computing (SBAC-PAD), Lyon, France* (2018), pp. 25–28

30. X. Li, V. Kashyap, J.K. Oberg, M. Tiwari, V.R. Rajarathinam, R. Kastner, T. Sherwood, B. Hardekopf, Ben, F.T. Chong, Sapper: a language for hardware-level security policy enforcement. ACM SIGPLAN Not. **49**(4), 97–112 (2014)

31. M. Lipp, M. Schwarz, D. Gruss, T. Prescher, W. Haas, A. Fogh, J. Horn, S. Mangard, P. Kocher, D. Genkin, Y. Yarom, M. Hamburg, Meltdown: reading kernel memory from user space, in *27th Security Symposium (USENIX Security)* (IEEE, Piscataway, 2018), pp. 973–990

32. Y. Lyu, X. Qin, M. Chen, P. Mishra, Directed test generation for validation of cache coherence protocols, in *IEEE Transactions on Computer-Aided Design of Integrated Circuits and Systems (TCAD)*, February 2018 (IEEE, Piscataway, 2018)

33. Y. Lyu, A. Ahmed, P. Mishra, Automated activation of multiple targets in RTL models using Concolic testing, in *Design Automation and Test in Europe (DATE)* (IEEE, Piscataway, 2019)

34. P. Mishra, M. Chen, Efficient techniques for directed test generation using incremental satisfiability, in *International Conference on VLSI Design* (2009), pp. 65–70

35. P. Mishra, N. Dutt, Graph-based functional test program generation for pipelined processors, in *Design Automation and Test in Europe (DATE)* (IEEE, Piscataway, 2004), pp. 182–187

36. P. Mishra, N. Dutt, Functional coverage driven test generation for validation of pipelined processors, in *Design Automation and Test in Europe (DATE)* (IEEE, Piscataway, 2005), pp. 678–683

37. P. Mishra, N. Dutt, Specification-driven directed test generation for validation of pipelined processors. ACM Trans. Des. Autom. Electron. Syst. (TODAES) **13**(2), 36, article 42 (2008)
38. A. Nahiyan, K. Xiao, K. Yang, Y. Jin, D. Forte, M. Tehranipoor, AVFSM: a framework for identifying and mitigating vulnerabilities in FSMS, in *2016 53nd ACM/EDAC/IEEE Design Automation Conference (DAC)* (IEEE, Piscataway, 2016), pp. 1–6
39. A. Nahiyan, F. Farahmandi, D. Forte, P. Mishra, M. Tehranipoor, Security-aware FSM design flow for identifying and mitigating vulnerabilities to fault attacks, in *IEEE Transactions on Computer-Aided Design of Integrated Circuits and Systems (TCAD)* (IEEE, Piscataway, 2018)
40. A. Pnueli, The temporal logic of programs, in *18th Annual Symposium on Foundations of Computer Science (SFCS 1977)* (IEEE, Piscataway, 1977), pp. 46–57
41. S. Proch, P. Mishra, Test generation for hybrid systems using clustering and learning techniques, in *International Conference on VLSI Design* (2016), pp. 589–590
42. X. Qin, P. Mishra, Directed test generation for validation of multicore architectures. ACM Trans. Des. Autom. Electron. Syst. (TODAES) **17**(3), article 24, 21 (2012)
43. X. Qin, P. Mishra, Automated generation of directed tests for transition coverage in cache coherence protocols, in *Design Automation and Test in Europe (DATE)* (IEEE, Piscataway, 2012)
44. X. Qin, P. Mishra, Scalable test generation by interleaving concrete and symbolic execution, in *International Conference on VLSI Design* (IEEE, Piscataway, 2014), pp. 104–109
45. X. Qin, M. Chen, P. Mishra, Synchronized generation of directed tests using satisfiability solving, in *International Conference on VLSI Design* (2010), pp. 351–356
46. Y. Qiu, H. Li, T. Wang, B. Liu, Y. Gao, X. Li, Property coverage analysis based trustworthiness verification for potential threats from eda tools, in *2016 IEEE 25th Asian Test Symposium (ATS)* (IEEE, Piscataway, 2016), pp. 43–48
47. J. Rajendran, V. Vedula, R. Karri, Detecting malicious modifications of data in third-party intellectual property cores, in *Proceedings of the 52nd Annual Design Automation Conference* (ACM, New York, 2015), pp. 1–6
48. J. Rajendran, A.M. Dhandayuthapany, V. Vedula, R. Karri, Formal security verification of third party intellectual property cores for information leakage, in *2016 29th International Conference on VLSI Design and 2016 15th International Conference on Embedded Systems (VLSID)* (IEEE, Piscataway, 2016), pp. 547–552
49. M. Rathmair, F. Schupfer, Hardware trojan detection by specifying malicious circuit properties, in *2013 IEEE 4th International Conference on Electronics Information and Emergency Communication (ICEIEC)* (IEEE, Piscataway, 2013), pp. 317–320
50. M. Tehranipoor, F. Koushanfar, A Survey of Hardware Trojan Taxonomy and Detection. IEEE Des. Test Comput. **27**(1), pp. 10–25 (2010)
51. M. Tiwari, H.M.G. Wassel, B. Mazloom, S. Mysore, F.T. Chong, T. Sherwood, Complete information flow tracking from the gates up. ACM Sigplan Not. **44**(3), 109–120 (2009)
52. Tutorial on SystemVerilog Assertions. https://www.project-veripage.com/sva_1.php
53. S. Vasudevan, E.A. Emerson, J.A. Abraham, Efficient model checking of hardware using conditioned slicing, Electron. Notes Theor. Comput. Sci. **128**(6), 279–294 (2005)
54. S. Vijayaraghavan, M. Ramanathan, *A Practical Guide for SystemVerilog Assertions* (Springer Science and Business Media, Berlin, 2005)
55. D. Zhang, Y. Wang, G.E. Suh, A.C. Myers, A hardware design language for timing-sensitive information-flow security. ACM SIGARCH Comput. Arch. News **43**(1), 501–516 (2015)

Chapter 8
Automated Test Generation for Detection of Malicious Functionality

8.1 Introduction

A Trojan is expected to be covert and difficult to detect, i.e. an intelligent adversary will likely insert a Trojan circuit in a way that evades detection during post-manufacturing functional/parametric testing, but manifests itself during long hour of in-field operation. This can be achieved by externally triggering its operation or by making it dependent on rare circuit conditions inside an IC. The condition of Trojan activation is commonly referred to as *trigger condition*, which can be purely combinational or sequential, i.e. related to the clock or a sequence of rare events in the state elements (e.g., flip-flops of registers). The internal circuit nodes affected by a Trojan activation are referred to as *payload* of a Trojan. Figure 8.1 shows some example Trojan circuits including a combinational and a sequential Trojan. For example, a Trojan circuit could be triggered only when a data bus attains a unique rare value or when the number of times it attains the rare value equals to a particular count. The malicious effects of Trojan payloads can range from passive, such as leakage of secret information to altering the original functionality of the chip in a critical or destructive fashion.

The underlying assumption for Trojan insertion is that an adversary is fully aware of the design functionality and therefore can hide the Trojan in a hard-to-find place. The adversary may use very rare internal transitions to trigger the Trojan, and it may be impossible to detect (due to exponential state space) during traditional testing and validation. The major challenges for generating high-quality test vectors are as follows: (1) we are not sure of the location where the Trojan is inserted in the circuit; (2) the Trojan is stealthy and has very low activity when it is not triggered. These characteristics have made random tests not effective in magnifying the side-channel signal for Trojan detection. Figure 8.2 shows two example Trojan instances. The 4-trigger Trojan will only be activated by the rare combination 1011 and the 8-trigger Trojan will only be activated by the rare combination 10110011. If the possibility of

© Springer Nature Switzerland AG 2020
F. Farahmandi et al., *System-on-Chip Security*,
https://doi.org/10.1007/978-3-030-30596-3_8

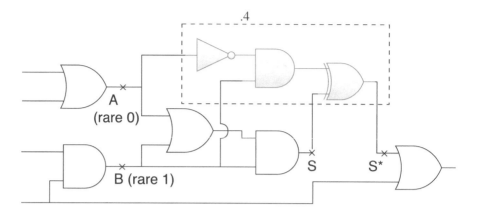

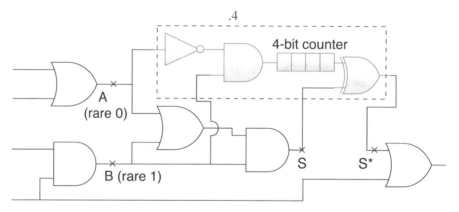

Fig. 8.1 Example of a combinational and a sequential Trojan with triggers from two rare internal nodes A and B. (**a**) Combinational Trojan. (**b**) Sequential Trojan

each rare node to take its rare value is 0.1, the probability to have these two Trojans fully triggered is 10^{-4} and 10^{-8}, respectively.

To detect a hardware Trojan with testing, we need to (1) trigger the rare activation conditions and (2) propagate the effects to observable outputs. Trojans have been carefully crafted and implanted so that they can remain dormant for almost all the functionality testing and manufacturing testing. Even in some cases the Trojan is activated, its effects might not be able to propagate to observable outputs and thus it will not be detected. These special characteristics of hardware Trojans have posed great challenges for test generation strategies to detect them. We will describe several approaches that focus on test generation in order to face these challenges in the rest of this chapter.

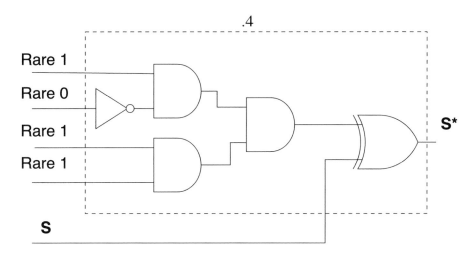

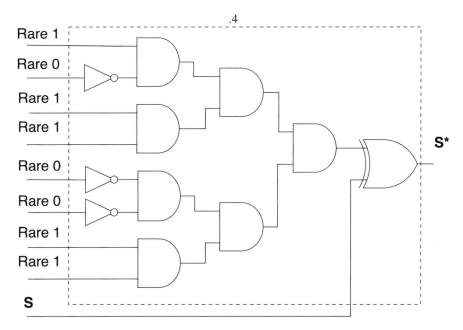

Fig. 8.2 Trojans with rare nodes as trigger conditions. The 4-trigger Trojan will only be activated by the rare combination 1011 and the 8-trigger Trojan will only be activated by the rare combination 10110011. (**a**) A 4-trigger Trojan. (**b**) An 8-trigger Trojan

8.2   Mutation-Based Random Test Generation

A major problem with formal methods and ATPG based Trojan detection methods is the scalability issue. Formal methods face the state space explosion issue when design is large. ATPG can be used to activate a Trojan if all the triggers are known. However, this is not feasible for Trojan detection since Trojans are likely to have unknown number of triggers hidden at stealthy locations. It would be practically infeasible to use ATPG to test all possible trigger conditions. MERO [4] takes the advantage of N-detect test (see [21] of Chap. 10) to maximize the trigger coverage by activating the rare nodes. The test generation ensures that each of the nodes gets activated to their rare values for at least N times. It is shown that if N is sufficiently large, a Trojan with trigger condition based on these rare nodes is likely to be activated by the generated test set. Saha et al. [36] improve the test pattern generation of MERO [4] by using genetic algorithm and Boolean satisfiability for ATPG. Their approach could more effectively propagate the payload of possible Trojan candidates. Since MERO is the most prominent in mutation-based random test generation for hardware Trojan detection, we explain the detailed steps of MERO in this section.

Figure 8.3 shows the flowchart for MERO methodology. Given a circuit netlist and a set of parameters, it first determines the rare nodes using the RO-Finder

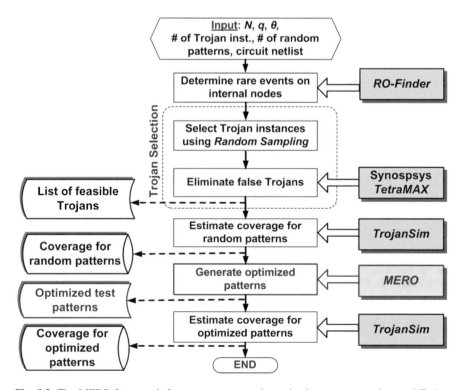

Fig. 8.3 The MERO framework for rare occurrence determination, test generation, and Trojan simulation [4]

(rare occurrence finder). These rare nodes will be the trigger points where potential Trojans embed themselves in the circuit. For example, a q-trigger random Trojan instance is created by randomly selecting the trigger nodes from the list of rare nodes. The Trojan trigger conditions will be justified with TetraMAX to eliminate false Trojans. Then random test patterns are generated. Their MERO algorithm will work on top of random test patterns, and turn them into a high-quality test set by mutating the random test vectors to achieve good coverage on rare nodes. Finally MERO patterns and random patterns are compared in terms of trigger coverage and Trojan coverage.

The MERO algorithm works as follows. They start with the golden circuit netlist (without any Trojan), a list of rare nodes (L), and the number of times to activate each node to its rare value (N). First, a random pattern set (V) is generated as the seeds. For each rare node in L, they count the number of times that it encounters a rare value. For each random pattern v_i in V, they count the number of nodes (C_R) whose rare value is satisfied, and sort vectors in decreasing order of C_R. In the next step, they mutate one bit at a time for each seed vector. Mutated vector is accepted if it can increase the number of nodes satisfying their rare values. The process repeats until each node in L satisfies its rare value at least N times. The output test patterns will be a minimal test set that improves the coverage compared to random patterns.

Figure 8.4 shows the trigger and Trojan coverage for a ISCAS-85 benchmark. Along with the increasing values of N, we can clearly see that both the test length and the quality of test set are improving. The trigger coverage and Trojan coverage obtained with the MERO approach increase steadily with N. This trend is similar to $N - detect$ tests for stuck-at-fault where defect coverage improves with increasing N.

Table 8.1 lists the detailed results for a set of combinational and sequential benchmarks from the ISCAS benchmarks. The trigger and Trojan coverage for

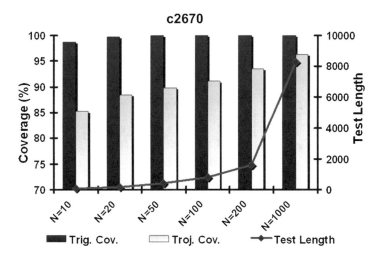

Fig. 8.4 Trigger coverage and Trojan coverage of MERO [4]

Table 8.1 Comparison of Trigger and Trojan coverage among ATPG, Random, and MERO patterns [4]

Circuit	Nodes (rare/total)	ATPG patterns				Random (100 K patterns)				MERO patterns			
		q = 2		q = 4		q = 2		q = 4		q = 2		q = 4	
		Trig. Cov. (%)	Troj. Cov. (%)	Trig. Cov. (%)	Troj. Cov. (%)	Trig. Cov. (%)	Troj. Cov. (%)	Trig. Cov. (%)	Troj. Cov. (%)	Trig. Cov. (%)	Troj. Cov. (%)	Trig. Cov. (%)	Troj. Cov. (%)
c2670	297/1010	93.97	58.38	30.7	10.48	98.66	53.81	92.56	30.32	100	96.33	99.9	90.17
c3540	580/1184	77.87	52.09	16.07	8.78	99.61	86.5	90.46	69.48	99.81	86.14	87.34	64.88
c5315	817/2485	92.06	63.42	19.82	8.75	99.97	93.58	98.08	79.24	99.99	93.83	99.06	78.83
c6288	199/2448	55.16	50.32	3.28	2.92	100	98.95	99.91	97.81	100	98.94	92.5	89.88
c7552	1101/3720	82.92	66.59	20.14	11.72	98.25	94.69	91.83	83.45	99.38	96.01	95.01	84.47
s13207	865/2504	82.41	73.84	27.78	27.78	100	95.37	88.89	83.33	100	94.68	94.44	88.89
s15850	959/3004	25.06	20.46	3.8	2.53	94.2	88.75	48.1	37.98	95.91	92.41	79.75	68.35
s35932	970/6500	87.06	79.99	35.9	33.97	100	93.56	100	96.8	100	93.56	100	96.8
Avg.	724/2857	74.56	58.14	19.69	13.37	98.84	88.15	88.73	72.3	99.39	93.99	93.5	82.78

stuck-at ATPG patterns, weighted random patterns, and MERO test patterns are compared. It also shows the number of rare nodes and total nodes in each circuit. Initially the signal probabilities were simulated with a set of 100,000 random vectors. For combinational circuits, 100,000 random Trojan instances are simulated. For sequential circuits, 10,000 random Trojan instances are simulated to reduce the runtime of TetraMax. The parameters for MERO include the number of rare occurrences ($N = 1000$), the number of Trojan triggers ($q = 2$ and $q = 4$), and rare probability threshold ($\theta = 0.2$). The ATPG patterns provide poor coverage results compared to MERO patterns, and it is more obvious in the case of higher number of trigger pointers. We can also see that Trojan coverage is consistently smaller compared to trigger coverage. For a Trojan to be covered, we do not only need all of its triggers to be covered, but also need its payload to be propagated to observable primary outputs. In many cases, the trigger condition is satisfied, but the malicious effect is not propagated to outputs and the Trojan is undetected.

As the reader might have realized, MERO has its shortcomings. (1) MERO starts with a random set of 100,000 patterns. It is a fairly small search space to mutate one bit at a time to generate new vectors. (2) For Trojans with very rare triggers (for example, very "hard-to-trigger" Trojans with triggering probability in the range of 10^{-6} or less), the test vectors generated by MERO were found to have poor coverage [36]. It was found that best coverage was achieved for θ in the range 0.08-0.12. (3) Another problem with MERO algorithm is that the payload of Trojan might not get propagated to any primary output. In other words, MERO focuses only on the triggers of a Trojan while it ignores the propagation of payload effects.

Several improvements to MERO has been proposed to further improve coverage as well as its application in side-channel aware test generation. Saha et. al. [36] improved MERO by boosting the test generation with genetic algorithm and SAT-based approach. This can greatly relieve the first two limitations mentioned above. They also made the test generator to select test patterns which are payload-aware for the third limitation. Huang et. al. [20, 21] proposed test generation for side-channel analysis based Trojan detection. They targeted on taking advantage of high-quality test generation to be side-channel aware, and the generated tests will magnify the side-channel signals for Trojan detection.

8.3 Directed Test Generation Using Formal Methods

For any bug/defect/Trojan to be detected, the quality of the test vectors matters. If we already know where the location of the Trojan is, we can generate directed tests which target on the specific location. The stealthy feature of Trojans implies that they are usually implanted at hard-to-trigger locations. If we can get these rare locations, directed test generation can help us in detecting Trojans.

Directed tests are carefully designed to check particular behaviors of the design. They are very promising in reducing the overall validation effort since a drastically small amount of tests are required compared to random tests to obtain the same coverage goal. Directed test generation is mostly performed by human intervention. Formal methods like SAT-based approach and model checking approach can help with directed test generation. Automatic test pattern generation (ATPG) can also help with targeted test generation. In this section, we will describe these directed test generation approaches in detail.

8.3.1   Test Generation Using SAT Solvers

Given a Boolean formula, the satisfiability problem relies on finding Boolean values to the formula's variables such that the formula is evaluated to true. If such an assignment does not exist, the formula is called unsatisfiable. It means that any possible assignments to formula's variables force the formula to be evaluated to false. The Boolean formula is constructed from AND, OR, and NOT operators between various variables which can be either assigned to true or false. Many of the validation and debugging problems can be mapped to satisfiability problems. One of the applications is to check the equivalence between the specification of the circuit and its implementation using SAT solvers. Figure 8.5 shows the equivalence checking using SAT solvers. If the specification and implementation have the same functionality, the output of the XOR gate should always be false. If the output of XOR gate becomes true for any input pattern, it implies that the implementation and the specification do not have the same functionality for the same input pattern. In other words, if the circuit shown in Fig. 8.5 is converted to conjunction normal form (CNF), a SAT solver can be used to check the equivalence between the specification and implementation. If the SAT solver reports unsatisfiable, we can conclude that specification and implementation are equivalent. Otherwise, they are not equivalent and the root of mismatch should be found.

Equivalence checking can be done using SAT solvers to identify hardware Trojans [19]. If hardware Trojans exist in the implementation, the SAT solver finds assignments to the internal variables to reveal the hidden Trojan. However, this method requires a golden model and suffers from scalability issues. The SAT solver may encounter state explosion when the design is large, and the specification and the implementation significantly differ from each other.

Several works explore the existence of Trojans in unspecified functionality [16, 17]. Therefore, the Trojan does not alter the specification of the design, and existing statistical or simulation-based methods cannot identify the Trojan-inserted design [18]. Fern et al. propose a SAT-based technique to detect Trojans which exploit

Fig. 8.5 Equivalence checking using SAT solvers

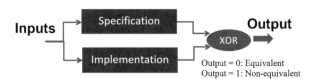

the design signals in their unspecified functionality to cause malfunction. Fern et al. try to address unspecified Trojan detection where the Trojan targets information leakage [18]. Suppose that the function "$func$" is unspecified when internal signal "s" is under condition "C". Suppose that signal s can have two possible values: v_0 and v_1. Under condition C, Eq. 8.1 should be unsatisfiable if the design is Trojan-free. Therefore, any assignment which makes Eq. 8.1 satisfiable is a trace to detect the covert Trojan.

$$C \wedge (func(s = v_0) \oplus func(s = v_1)) \tag{8.1}$$

To detect Trojans in an unspecified functionality of the design, pairs C and s should be identified. For any function in the design, several s and C pair can be found, and the process of marking the potential pairs is not automatic yet. For every pair (s, C), one CNF formula is constructed and an SAT solver (for Boolean values) or a Satisfiability Module Theory solvers (SMT-solvers) can be used to find the potential threats. The Trojan can be detected when the CNF formula is satisfiable. The success of this approach is dependent on the SAT solvers and identifying (s, C) pairs. Moreover, the approach requires manual intervention.

8.3.2 Test Generation Using Model Checking

Model checking is a famous technique in design verification which checks a design for a set of given properties. A model checker tries all of the possible states of a design to prove a given property using a binary decision diagram (BDD). However, the number of design states can be huge since every bit introduces two states in the design. For example, a 32-bit register can add 2^{32} states to the design state space. Although some techniques such as slicing, abstracting, etc. have been proposed [10, 37], state space explosion still is the largest limitation of using model checking in property verification. Bounded model checking (BMC) is introduced to overcome the amount of memory that a model checker requires for constructing and storing different states of a design [3]. BMC tries to find a counter-example in the first K cycles during execution. If a counter-example is found within K cycles, the property does not hold. Otherwise, K can be increased in the hope of finding a counter-example in upper bounds. BMC is not able to prove a property since it unrolls the circuit for a specified number of clock cycles. However, it can provide a statistical metric for a given property when the model checker fails (e.g., no counter-example can be found in K clock cycles). The BMC problem can be mapped to satisfiability problem, and SAT solvers can be utilized to solve the problem. Therefore, the BMC addresses some of the state space explosion problems associated with BDDs in model checking. Figure 8.6 shows the bounded model checking approaches, respectively. Clearly, bounded model checking cannot provide proof for property P. However, it can reveal when property P is violated within K clock

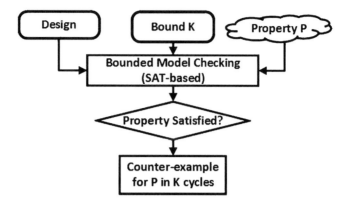

Fig. 8.6 Verification using bounded model checking

cycles. There is a vast literature in directed test generation using SAT-based BMC [5–9, 13–15, 23–25, 30–32].

As described in Chap. 7, security properties describe the expected behaviors which a trustworthy design is required to follow. Model checkers can be used to ensure safety properties. A SoC designer and a third-party vendor can agree on certain security properties that should be held on the design. When the design is sent to the SoC integrator, the SoC integrator converts the design to a formal description to check the security properties using a model checker. If all of the security properties are verified, the expected security behaviors are met. Rajendran et al. have proposed a Trojan detection technique which is based on using bounded model checking [35]. They have considered the threat model as an attempt to corrupt the critical data such as secret keys of a cryptographic design, random numbers which are required by most of the cryptography algorithms, or stack pointer of a processor. The assumption is that these critical data should be stored in some specific registers and accesses to these registers should be protected. In other words, the registers which contain critical data should be accessed through valid ways, and any undefined access to these registers is considered as a threat. The safe access conditions to these registers are formulated as properties (assertions), and a bounded model checker is utilized to find a counter-example when the security properties are violated.

8.3.3 ATPG for Trojan Detection

ATPG has been widely used in chip design and manufacturing to detect defects in circuits. It searches for test patterns which can distinguish between the correct circuit behavior and the faulty circuit behavior caused by defects. The generated patterns are used to detect such defects and assist with failure analysis for root-cause of the defects. Metrics to evaluate the effectiveness of ATPG include the number of modeled faults (fault coverage) and the test application time. ATPG efficiency is influenced by the fault model (the stuck-at faults, transistor faults, bridging faults, opens faults, or delay faults etc.), the type of circuit under test (full scan, synchronous sequential, or asynchronous sequential), and the level of abstraction used to represent the circuit under test (gate, register-transfer, switch). Even for a pattern set with 100% fault coverage for a certain fault model, it does not mean it can detect all possible defects in the circuit. So as it comes to Trojan, it cannot guarantee that there is no Trojan in circuit. A well-crafted Trojan might still hide itself through ATPG test patterns. But still, it would help us gain some basic level of trust of circuit in the battle against Trojans.

ATPG can also help generate test patterns for the formal methods for Trojan detection. As for the verification using BMC in Fig. 8.6, the success of BMC is dependent on the SAT-engine (it may fail for large and complex designs) and precise definition of security properties which needs prior knowledge of all safe ways to access a critical register. The performance of the presented method can be improved using ATPG to ensure the trustworthiness of the assets for a large number of clock cycles. In the next section we will discuss such a hybrid approach which combines the advantage of both model checking and ATPG.

8.4   Test Generation Using ATPG and Model Checking

ATPG works well on full-scan designs, whereas model checking is suitable for logic blocks without scan chain. Due to overhead considerations, partial-scan chain insertion is the standard practice today. Unfortunately, neither ATPG nor model checking is suitable for partial-scan designs. We propose a hybrid approach [11, 12, 22] which combines ATPG and model checking. We use model checking on the subset of non-scan elements and ATPG on the scan elements to avoid common pitfalls of running the original design. Experimental results demonstrate the effectiveness of tests generated for Trojan detection on Trust-hub benchmarks.

As shown in Fig. 8.7, this approach identifies suspicious branches/gates which may be used as triggering conditions for hardware Trojans. In order to generate tests to activate rare nodes, scan replacement is done in the next step. We generate security properties that targets activation of equivalent signals/gates of rare nodes in the gate-level netlist. The scan replaced netlist as well as the security properties are used by the model checker. We generate a set of constraints using model checker to facilitate directed test generation using ATPG tool.

Algorithm 17 shows the major steps in the framework shown in Fig. 8.7. Algorithm 17 takes a design D, and outputs a set of test vectors T. The set of rare nodes R is identified in the design which are used by the constraint Generation and test Vector Generation procedures. The constraint Generation procedure uses model checking to produce a set of signal traces S. Finally the test Vector Generation method uses ATPG with the design, rare nodes, and signal traces to produce a set of test vectors for activating each rare node.

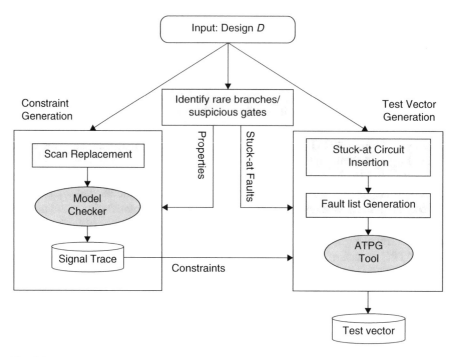

Fig. 8.7 The overview of test generation using model checking and ATPG [22]

Algorithm 17: Trojan detection using ATPG

Input: Design D
Output: Set of testvectors T
procedure TROJAN DETECTION(D)
R,S,T =
/* Step 1: Identify suspicious points */
R = identifyRareBranches(D)
/* Step 2: Model checking */
S = constraintGeneration(D,R)
/* Step 3: ATPG test generation */
T = testVectorGeneration(D,R,S)
return *T*

Rare Branch Identification   For each IP, initial analysis is performed at the RTL level to determine suspicious gates in the design. In a design, rare branches are branches that are not covered after running random tests up to millions of cycles. Mapping the RTL branches to gate-level netlist after synthesis is done in two phases. The first phase identifies any suspicious boundary and register nets and uses the synthesis tool commands to attempt to preserve suspicious signal nets. In these cases, identifiable naming will be preserved after synthesis. If any rare branch is not accounted for, then, the second phase constructs a structural dependency graph of the two representations and attempts to match these graphs using approximate

Benchmarks	Scan FFs (Scan/Total)	Test Cov.	#Rare Bran.	ATPG		Model Chk. (MC)		MERO		ATPG+MC	
				Detect	Time	Detect	Time	Detect	Time	Detect	Time
AES-T1000	6448/6933	99%	2	✓	0.02s	✓	85.6s	X	TO	✓	8.80s
AES-T2000	6468/7108	99%	5	✓	0.90s	✓	216.5s	X	TO	✓	22.0s
RS232-T400	30/59	97%	2	✓	0.24s	✓	3600s	✓	2810s	✓	0.52s
RS232-T800	26/58	97%	1	✓	0.06s	✓	7.23s	✓	3157s	✓	0.12s
cb_aes_15	5006/5889	99%	1	✓	28800s	X	MO	X	15720s	✓	7.85s
cb_aes_20	7262/7809	99%	1	✓	28800s	X	MO	X	16740s	✓	38.3s

Fig. 8.8 Comparison with ATPG, model checking and MERO [22]

graph matching heuristics. Other statistical or functional methods for determining rare nodes at RTL or gate-level such as FANCI [38] and MERS [20] are equally applicable.

Constraint Generation For each of the rare nodes identified, this step generates a set of signal traces to be used in ATPG. The algorithm takes the design, D, and replaces the scan FFs with pseudo-primary inputs. A property specified as the negation of the activation is generated for each rare node $r \in R$. Model checker then outputs a signal trace for each property.

Test Vector Generation The activation levels of all relevant internal signals from the suspicious nodes fan-in cone and scan replacements are extracted from the trace (from previous step) and combined together with an ATPG primitive AND gate referred to as stuck-at circuit. The addition of these primitives are for test generation purposes only and have no effect on the design functionality. A stuck-at 0 fault is added to the tools fault list for each stuck-at circuit. The ATPG tool is then run using full-sequential ATPG to generate test vectors that trigger each fault.

Figure 8.8 shows the comparison of this approach against ATPG, model checking, and MERO for Trojan detection. We can see the ATPG tool took a significant amount of time in generating test patterns to trigger the rare branch. Similarly model checking fails to generate a pattern due to state explosion and the tool experiencing a memory overflow (MO). The hybrid technique is able to generate test vectors even when the circuit structure has sufficient non-scan FF depth and structure. The results show that we can achieve up to four orders of magnitude faster execution times than state-of-the-art methods. This speed up is achieved by leveraging the strengths of each tool. Specifically, reducing the state space of model checking and removing the non-scan sequential complexity encountered using ATPG.

8.5 Scalable Test Generation Using Concolic Testing

Concolic testing generates tests by effective combination of concrete simulation and symbolic execution. Depending on the objective of the test generation, concolic testing can maximize coverage by forcing execution through different branches or can guide the execution towards a specific branch. It does so by alternating between concrete simulation and symbolic execution of the design. The first step involves the simulation of the design. For initial simulation, usually random inputs are used. The execution path taken by the simulation can be decomposed into a set of constraints, referred as path constraints ($C =< c_1, c_2, \ldots, c_n >$). Next step is to force the execution through an alternate branch. In order to do so, constraint of the selected branch (c_k) is negated and the desired path constraints for this alternate path is formed ($C' =< c_1, c_2, \ldots, \neg c_k >$). These new path constraints are then symbolically solved using a constraint solver. If the solver comes up with a solution input set, then for that input execution will go through this alternate branch. If no solution is found, another branch is selected for negation. These steps are repeated until required target branch is reached, or there is no solvable branch left. Other termination criteria such as timeout or coverage goal can also be used. Concolic testing avoids state explosion issue by exploring only one path at a time, instead of trying to explore all possible paths at once. However, these concolic testing methods only consider sequential execution models and are not applicable on hardware (concurrent) designs.

In this section, we propose a concolic testing based directed test generation approach for RTL designs [1, 2, 26–29, 33, 34]. This method utilizes distance feedback to quickly reach the desired security targets. As shown in Fig. 8.9, it consists of three major steps: (1) design instrumentation, (2) obtaining security targets of the design based on the identification of the rare branches and assignments, and (3) directed test generation to activate the security targets. The remainder of this section describes these steps in detail.

Design Instrumentation is needed to trace the execution paths during simulation of the design. Instrumentation is done by inserting a *display* statement for each functional statement. This insertion is automated and done during the abstract syntax tree (AST) generation phase of the simulator. Note that the instrumentation will not change the functionality of the design, since it only traces the executed statements. This trace is later used to identify rare branches as well as to generate path constraints for symbolic execution. The design needs to be instrumented only once.

Rare Branch Identification is achieved through random simulation to find rare branches which can potentially host hardware Trojans. We simulate the instrumented design using random inputs. Next, the number of times each branch is covered is counted. The branches that are covered less than a threshold number of times are marked as suspicious branches. For example, having a threshold of zero implies only uncovered branches as suspicious. The experiments uses a threshold of zero. It gives the lowest probability of false positive. All the branches that fall within the threshold are considered as security targets for the proposed Trojan activation framework.

Coverage Guided Test Generation is illustrated in Fig. 8.10. The algorithm takes an RTL design as well as security targets as inputs and generates directed tests to cover the security targets. First, we perform a preprocessing step to reduce the total number of security targets. The number of targets can be reduced based on the dependency between them due to the fact that all branches within a rare branch are

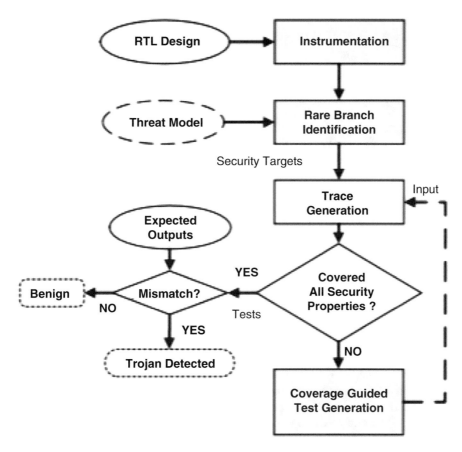

Fig. 8.9 Test geneartion framework using Concolic testing [2]

also rare. Covering the inside branch will also cover the parent branch, and thus it
can be removed from the target list. Such dependency can be resolved by looking
at the control flow graph (CFG) of the design. If a target is dominator of any other
target, it can be pruned. Here (Fig. 8.10a) shows the initial targets as B, D, and E.
However, B is a dominator of target D, hence can be removed. This is done statically,
without unrolling the design for multiple cycles. The static analysis only prunes part
of the dependent branches. Dynamic pruning with actual unrolling of design would
result in more pruned targets, but we do not use it in this work since it is susceptible
to state explosion.

After pruning step, one of the targets is selected for test generation. Distance from
the target is then evaluated by running breadth-first search (BFS) starting from the
target branch, and following predecessor edges in the CFG. An example is shown
in Fig. 8.10d. Here, D is selected to be covered first. Initially, target D is assigned
distance 0 and all other branches are assigned infinity. Next, we run BFS starting

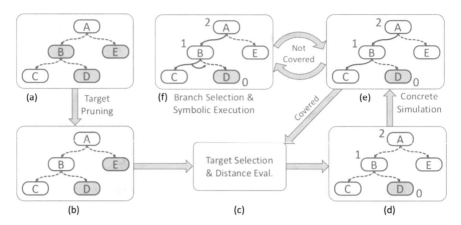

Fig. 8.10 Concolic test generation procedure [2]. Targets are shaded. (**a**) Initial targets. (**b**) Targets after pruning. (**c**), (**d**) Selects one target, and evaluates distance for that target. (**e**) Runs concrete simulation. Execution path is marked as red (solid line). (**f**) Selects an alternate branch and symbolically solves for input

from D, and follow predecessor edge. After distance evaluation is finished, the distance would be: $B = 1$, $A = 2$ and others infinity. This procedure is also done statically without actually unrolling the design. Next, we apply concrete simulation followed by symbolic execution for several iterations in order to generate tests to activate the potential hardware Trojan. In each iteration, the instrumented design is simulated for a specific number of clock cycles (i.e., unroll cycles) and a trace file is produced (Fig. 8.10e). The information of the trace file is then converted into path constraints. These constraints model the execution path taken by the concrete simulation. In the next step, one of the alternate branches is selected to be explored. We have selected the branch which has lowest assigned distance value. In other words, we have given priority to the branches that are closer to our security target. Path constraints that lead to that branch is then symbolically solved by a constraint solver. If a solution exists, then we again do concrete simulation with that solution, this time forcing execution through that alternate branch. This concrete and symbolic execution steps are repeated until the target is covered (Fig. 8.10e–f), or some terminating conditions are met (e.g., timeout). If all of the branches are exhausted and no new input vector can be generated, algorithm returns generated tests.

Intuitively, the iterative procedure effectively guides the execution path towards the target. This approach avoids the state space explosion by examining one path at a time in contrast to traditional formal methods that consider all of the paths simultaneously. Therefore, it is capable of activating hard-to-detect Trojans in large designs, as will be demonstrated in the experiments section.

8.6 Summary

In this chapter, we focus on different test generation approaches for hardware Trojan detection. Test generation is widely used for detecting bugs and helping us analyze defects in circuit. When we apply test generation for Trojan detection, we face the challenge that Trojans are stealthy in nature and can hide itself through most common functional tests. Special care must be taken for test generation to be effective for Trojan detection. We explain mutation-based random test generation approaches like MERO. We explain the application of directed test generation which uses formal methods (like SAT solvers and model checking) or ATPG apporach to target on possible Trojan trigger conditions. Formal methods have limitations when the design is large and search space is huge. We introduce a latest approach which combines formal method (model checking) and traditional test generation (ATPG). We also introduce concolic testing for test generation in hardware Trojan detection. All these test generation approaches shed light on improving test coverage and establishing trustworthiness of designs against hardware Trojan attacks.

References

1. A. Ahmed, P. Mishra, QUEBS: qualifying event based search in concolic testing for validation of RTL models, in *IEEE International Conference on Computer Design (ICCD)*, pp. 185–192 (2017)
2. F.F. Alif Ahmed, P. Mishra, Directed test generation using concolic testing of rtl models, in *Design Automation and Test in Europe (DATE)* (IEEE, Piscataway, 2018), pp. 1538–1543
3. A. Biere, A. Cimatti, E.M. Clarke, O. Strichman, Y. Zhu, Bounded model checking. Adv. Comput. **58**, 117–148 (2003)
4. R. Chakraborty, F. Wolff, S. Paul, C. Papachristou, S. Bhunia, MERO: a statistical approach for hardware trojan detection, in *International Workshop on Cryptographic Hardware and Embedded Systems* (IEEE, Piscataway, 2009), pp. 396–410
5. M. Chen, P. Mishra, Functional test generation using efficient property clustering and learning techniques. IEEE Trans. Comput. Aided Des. Integr. Circuits Syst. (TCAD) **29**(3), 396–404 (2010)
6. M. Chen, P. Mishra, Decision ordering based property decomposition for functional test generation, in *Design Automation and Test in Europe (DATE)* (IEEE, Piscataway, 2011), pp 167–172
7. M. Chen, P. Mishra, Property learning techniques for efficient generation of directed tests. IEEE Trans. Comput. (TC) **60**(6), 852–864 (2011)
8. M. Chen, P. Mishra, D. Kalita, Automatic RTL test generation from SystemC TLM specifications. ACM Trans. Embed. Comput. Syst. (TECS) **11**(2), article 38 (2012)
9. M. Chen, X. Qin, P. Mishra, Learning-oriented property decomposition for automated generation of directed tests. Springer J. Electron. Test. (JETTA) **30**(3), 287–306 (2014)
10. A. Cimatti, E. Clarke, F. Giunchiglia, M. Roveri, NUSMV: a new symbolic model checker. Int. J. Softw. Tools Technol. Transfer **2**(4), 410–425 (2000)
11. J. Cruz, Y. Huang, P. Mishra, S. Bhunia, An automated configurable Trojan insertion framework for dynamic trust benchmarks, in *Design Automation and Test in Europe (DATE)* (IEEE, Piscataway, 2018)

12. J. Cruz, P. Mishra, S. Bhunia, The metric matters: how to measure trust, in *Design Automation Conference (DAC), Las Vegas*, 2–6 June 2019 (2019)
13. N. Dang, A. Roychoudhury, T. Mitra, P. Mishra, Generating test programs to cover pipeline interactions, in *ACM/IEEE Design Automation Conference (DAC)* (2009), pp. 142–147
14. F. Farahmandi, P. Mishra, Automated test generation for debugging arithmetic circuits, in *Design Automation and Test in Europe (DATE)* (IEEE, Piscataway, 2016), pp. 1351–1356
15. F. Farahmandi, P. Mishra, Automated test generation for debugging multiple bugs in arithmetic circuits. IEEE Trans. Comput. (TC) **68**(2), 182–197 (2019)
16. N. Fern, S. Kulkarni, K.-T.T. Cheng, Hardware trojans hidden in RTL don't cares—Automated insertion and prevention methodologies, in *Test Conference (ITC), 2015 IEEE International* (IEEE, Piscataway, 2015), pp. 1–8
17. N. Fern, I. San, C.K. Koç, K.-T.T. Cheng, Hardware trojans in incompletely specified on-chip bus systems, in *Proceedings of the 2016 Conference on Design, Automation and Test in Europe* (EDA Consortium, San Jose, 2016), pp. 527–530
18. N. Fern, I. San, K.-T.T. Cheng, Detecting hardware trojans in unspecified functionality through solving satisfiability problems, in *Design Automation Conference (ASP-DAC), 2017 22nd Asia and South Pacific* (IEEE, Piscataway, 2017), pp. 598–504
19. X. Guo, R.G. Dutta, Y. Jin, F. Farahmandi, P. Mishra, Pre-silicon security verification and validation: a formal perspective, in *ACM/IEEE Design Automation Conference (DAC)* (IEEE, Piscataway, 2015)
20. Y. Huang, S. Bhunia, P. Mishra, MERS: statistical test generation for side-channel analysis based trojan detection, in *Proceedings of the 2016 ACM SIGSAC Conference on Computer and Communications Security*, Series CCS New York, NY, USA (ACM, New York, 2016), pp. 130–141. [Online]. Available: http://doi.acm.org/10.1145/2976749.2978396
21. Y. Huang, S. Bhunia, P. Mishra, Scalable test generation for Trojan detection using side channel analysis, in IEEE Trans. Inf. Forensics Secur. **13**(11), 2746–2760 (2018)
22. A.A. Jonathan Cruz, F. Farahmandi, P. Mishra, Hardware trojan detection using atpg and model checking, in *International Conference on VLSI Design* (IEEE, Piscataway, 2018), pp. 91–96
23. H.-M. Koo, P. Mishra, Functional test generation using property decompositions for validation of pipelined processors, in *Design Automation and Test in Europe (DATE)* (IEEE, Piscataway, 2006), pp. 1240–1245
24. H.-M. Koo, P. Mishra, Functional test generation using design and property decomposition techniques. ACM Trans. Embed. Comput. Syst. (TECS) **8**(4), article 32 (2009)
25. Y. Lyu, X. Qin, M. Chen, P. Mishra, Directed test generation for validation of cache coherence protocols, in *IEEE Transactions on Computer-Aided Design of Integrated Circuits and Systems (TCAD)* (IEEE, Piscataway, 2018)
26. Y. Lyu, A. Ahmed, P. Mishra, Automated activation of multiple targets in RTL models using concolic testing, in *Design Automation and Test in Europe (DATE)* (IEEE, Piscataway, 2019)
27. P. Mishra, M. Chen, Efficient techniques for directed test generation using incremental satisfiability, in *International Conference on VLSI Design* (2009), pp. 65–70
28. P. Mishra, N. Dutt, Graph-based functional test program generation for pipelined processors, in *Design Automation and Test in Europe (DATE)* (IEEE, Piscataway, 2004), pp. 182–187
29. P. Mishra, N. Dutt, Functional coverage driven test generation for validation of pipelined processors, in *Design Automation and Test in Europe (DATE)* (IEEE, Piscataway, 2005), pp. 678–683
30. P. Mishra, N. Dutt, Specification-driven directed test generation for validation of pipelined processors. ACM Trans. Des. Autom. Electron. Syst. (TODAES) **13**(2), 36, article 42 (2008)
31. X. Qin, P. Mishra, Automated generation of directed tests for transition coverage in cache coherence protocols, in *Design Automation and Test in Europe (DATE)* (IEEE, Piscataway, 2012)
32. X. Qin, P. Mishra, Directed test generation for validation of multicore architectures. ACM Trans. Des. Autom. Electron. Syst. (TODAES) **17**(3), article 24, 21 (2012)
33. X. Qin, P. Mishra, Scalable test generation by interleaving concrete and symbolic execution, in *International Conference on VLSI Design* (IEEE, Piscataway, 2014), pp. 104–109

34. X. Qin, M. Chen, P. Mishra, Synchronized generation of directed tests using satisfiability solving, in *International Conference on VLSI Design* (2010), pp. 351–356
35. J. Rajendran, V. Vedula, R. Karri, Detecting malicious modifications of data in third-party intellectual property cores, in *ACM/IEEE Design Automation Conference (DAC)* (IEEE, Piscataway, 2015), pp. 112–118
36. S. Saha, R. Chakraborty, S. Nuthakki, Anshul, D. Mukhopadhyay, Improved test pattern generation for hardware trojan detection using genetic algorithm and boolean satisfiability, in *Cryptographic Hardware and Embedded Systems (CHES)* (Springer, Berlin, 2015), pp. 577–596
37. S. Vasudevan, E.A. Emerson, J.A. Abraham, Efficient model checking of hardware using conditioned slicing. Electron. Notes Theor. Comput. Sci. **128**(6), 279–294 (2005)
38. A. Waksman, M. Suozzo, S. Sethumadhavan, FANCI: Identification of stealthy malicious logic using boolean functional analysis, in *ACM SIGSAC Conference on Computer and Communications Security* (IEEE, Piscataway, 2013), pp. 697–708

Chapter 9
Trojan Detection Using Machine Learning

9.1 Introduction

Machine learning techniques have the capability to explore high-dimensional feature space and find patterns that are not intuitive for analytic approaches. To use machine learning techniques for hardware Trojan detection, we want machine learning's capability to extract relevant features and distinguish between Trojan-free and Trojan-infected designs. Machine learning can be tuned for hardware Trojan detection, almost in all aspects of hardware Trojan detection, i.e., logic testing [14], side-channel analysis [13, 22], functional/formal analysis [8–10, 30], runtime monitoring [5], etc.

For logic testing, machine learning can help generate test vectors that are more likely to have Trojans triggered or partially activated. In [29] genetic algorithm helps create test vectors that can active the Trojan. In [23] genetic algorithm is used for generating high-quality test vectors for maximum Trojan switching. For approaches based on structural or functional analysis, we can extract the structural or functional properties as features [11, 27] and train machine learning for classification. Approaches in [11, 27] can achieve more than 80% detection rate while they have high false positive rate. The same authors improve their feature selection and machine learning models in [12, 15] and manage to reduce false positive rate to below 10%. Insertion of ring oscillators into the circuit can provide features for genetic algorithm [18] or other machine learning approaches [19, 32] to accurately detect Trojans. For side-channel analysis, machine learning can build the pattern of side-channel fingerprints [16, 17, 26] of normal circuits and any outlier will be a Trojan circuit. These side-channel fingerprints rely on the availability of golden chips. Authors in [21] relieve this limitation by using fingerprints from process control monitors. For runtime Trojan detection or monitoring, machine learning can help as long as we can extract the runtime behavior into features and train the model properly. In [20], the authors implement machine learning into hardware for runtime monitoring of Trojans in the processor's routing network.

© Springer Nature Switzerland AG 2020
F. Farahmandi et al., *System-on-Chip Security*,
https://doi.org/10.1007/978-3-030-30596-3_9

This chapter discusses some of the successful cases of using machine learning in these aspects to inspire readers for more research in this exciting research domain. This chapter is organized as follows. First, we briefly explain the basics of commonly used machine learning techniques in Sect. 9.2. Next, we explain in detail cases of how machine learning is used for Trojan detection. Section 9.4 discusses in detail how genetic algorithm is used for test vector generation to active Trojans. Section 9.4 describes how we can generate features from gate-level netlists and train machine learning to classify circuits into Trojan-infected and Trojan-free ones. Section 9.5 shows an approach which uses ring oscillator network to create features for machine learning. Section 9.6 shows how one-class clustering can be used to generate boundary of side-channel fingerprints for Trojan detection. Section 9.7 discusses how machine learning is implemented on hardware to detect Trojans in routing network of many-core processors during runtime. Section 9.8 concludes the chapter.

9.2   Machine Learning Techniques

In this section, we explain some of the most popular machine learning techniques, which will be used in later sections for hardware Trojan detection. The reader who is familiar with machine learning can feel free to skip this section.

Support Vector Machine (SVM) is one of the most popular machine learning techniques for classification [25]. The training of SVM is to find the optimal hyperplane that separate data points with the largest margin possible between the two classes of data points. This hyperplane is the high-dimensional space which might be transformed from the feature space either linearly or by other kernel functions. The training of SVM involves selecting the kernel function (for example, linear, polynomial, radius basis function, etc.) and choosing the best parameters for the kernel. One-class SVM is a method which is based on the two-class SVM algorithm. As the name implies, it constructs the classification model based on only one class. It builds the separator hyperplane around the boundary of the training data. Hence, if any new data is different from the training data, it will fall outside the boundary and could be considered as outliers. Figure 9.1 illustrates the difference between two-class SVM and one-class SVM. As for hardware Trojan detection, we can use two-class SVM to classify a design as one of the two classes (i.e., Trojan-infected or Trojan-free), or use one-class SVM to classify a design as either Trojan-free or an outlier (Trojan-infected).

K-Nearest Neighbors (KNN) algorithm is a simple and effective machine learning method. It uses K training samples that are nearest to the new point in the feature space to vote for a majority decision. The new point will be classified as the same label as that of the majority. Thus for two-class classification, the value K is usually set as an odd number to prevent a split decision. The distance metric can be Euclidean distance or any other method of calculating distance. The distance metric and the value of K determine the accuracy of classification, as well as the

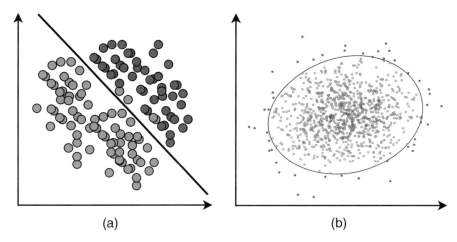

Fig. 9.1 Example of SVM hyperplanes: (**a**) two-class SVM finds the best hyperplane to separate data into two classes, (**b**) one-class SVM finds the best hyperplane to mark the boundary of training data

classification performance. Thus, it is important to search for an optimal K value for a data set by balancing the mean error and computation cost.

Genetic Algorithms (GA) are inspired by the process of natural selection [24] and widely used in computer science and operations research. They are commonly used to generate high-quality solutions for optimization and search problems. A typical setting of GA designs a vector to mimic the gene which usually has a very large search space that is not feasible for exhaustive search. A profit function defines the fitness of the vector (individual). By relying on bio-inspired operators such as mutation, crossover, and selection, GA searches the search space for high-fitness individuals or high-quality solutions. In each iteration (generation), the high-fitness individuals will be kept to continue the evolution. With generation by generation, GA drives the search towards solutions with higher and higher fitness. In test generation domain, genetic algorithm is shown to be successful in fault coverage [28] and Trojan detection [29].

The accuracy of classification algorithms depends on the data set, the features selected, the chosen machine learning model, and the parameters chosen for the model. The accuracy is quantified by these criteria: true positive rate (TPR), true negative rate (TNR), false positive rate (FPR), false negative rate (FNR). In the case of Trojan detection, true positive (TP) means the number of Trojans that get detected, true negative (TN) means the number of Trojan-free designs that are marked Trojan-free correctly, false positive (FP) means the number of Trojan-free designs that are classified as Trojan-infected mistakenly, false negative (FN) means the number of Trojan-infected designs that are classified as Trojan-free mistakenly. These criteria are calculated as follows:

$$\text{TPR} = \frac{\text{TP}}{\text{TP} + \text{FN}}$$

$$\text{TNR} = \frac{\text{TN}}{\text{TN} + \text{FP}}$$

$$\text{FPR} = 1 - \text{TNR} = \frac{\text{FP}}{\text{TN} + \text{FP}}$$

$$\text{FNR} = 1 - \text{TPR} = \frac{\text{FN}}{\text{TP} + \text{FN}}$$

For a good machine learning approach, we strive for high TPR and low FPR. High TPR means that we want as 100% of Trojan-infected designs to get detected if possible. Low FPR means that we do not want a Trojan-free design to be mis-classified as Trojan-infected, because each mis-classified design will require significant and unnecessary human effort to localize the Trojan.

9.3 GA-Based Test Generation for Trojan Detection

Logic testing in Trojan detection has been extensively explored, such as ATPG based [2, 6, 7, 31] and N-detect test [4]. The MERO approach presented in [4] utilized the idea of N-detect to achieve high coverage over randomly sampled Trojans, assuming the trigger conditions of the Trojans consist of rare nodes only. The authors observed that if the generated test patterns are able to satisfy all rare values N times, it is highly likely that rare trigger conditions are satisfied when N is sufficiently large.

Huang et al. [13] extended the idea of N-detect test for side-channel analysis, and proposed a test generation framework called MERS to maximize the sensitivity of dynamic current. MERS generates compact test patterns to let each rare node switch from its non-rare value to its rare value N times, increasing the probability of partially or fully activating a Trojan. The side-channel sensitivity of MERS is too small, typically less than 3% in most benchmarks [13], compared to large environmental noise and process variations in today's CMOS circuits.

As the difference of current switching in designs with/without Trojans comes from the inserted circuits and the switching after payloads are activated and propagated, the sensitivity can be improved if the test patterns can trigger rare conditions. Given any test pattern u generated in the previous step, the goal is to search for the best succeeding pattern v to maximize the sensitivity. There are three main challenges in searching for the best succeeding pattern for u. (1) Randomly selected pairs may not lead to high sensitivity, even if the two patterns are similar. For example, for a pair of random vectors, the current switching in G and G^T may remain the same, revealing no side-channel footprint. (2) The search space is exponentially large (2^n, where n is the number of inputs in the design). So, searching for the whole space is not feasible. (3) There is a tradeoff

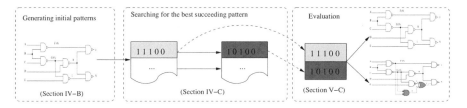

Fig. 9.2 Overview of genetic algorithm based test generation [23]

between introducing switching in the rare nodes and minimizing switching in the golden design. Figure 9.2 shows an overview of the proposed approach. It has three important steps. The first step finds the profitable initial test patterns. The next step forms the effective pair of test patterns. Finally, the quality of the generated pairs of test patterns needs to be evaluated. The remainder of this section briefly describes these steps.

Generation of Profitable Initial Test Patterns The sensitivity of side-channel analysis is maximized if the test pattern pairs are able to partially or fully activate trigger condition. Thus, the first task is similar to other logic testing techniques, such as ATPG or N-detect approach. We choose to use MERO [4] to generate N-detect test patterns. As introduced in [4], the generated test patterns are compact and can statistically achieve good coverage when N increases. MERO is used as a black box, and the parameters are introduced in Sect. 10.4. We denote the generated l test patterns as $\{u_i\}$ $(i = 1, 2, \ldots, l)$.

$$fitness_u(v) = \frac{rare\_switch^G_{u,v}}{switch^G_{u,v}} \tag{9.1}$$

Searching for the Best Succeeding Pattern Genetic algorithm consists of four major steps: initialization, fitness computation, selection, and crossover and mutation. The *fitness* is defined in Eq. 9.1, where $rare\_switch^G_{u,v}$ represents the current switching of all rare nodes in G when applying the test pattern u followed by v. A profitable test pattern should maximize the current switching in rare nodes to increase the probability of activating a Trojan, and minimize the switching in the golden design. The best succeeding pattern v_i for a given preceding u_i is the one achieving highest fitness value over all generations. The first iteration of GA for c17 is shown in Fig. 9.3, assuming four individuals in each generation.

The first step is *fitness computation*. For each individual v, the golden design G is simulated with the pair of test patterns (u_i, v). Then the fitness of v is computed by Eq. 9.1. For example, the fitness values for four candidates are shown in Fig. 9.3. *Selection* is based on the fitness of each individual. Individuals with higher fitness are more likely to be selected. The selection shown in Fig. 9.3 demonstrates that the individuals with higher fitness values (such as 10100) are more likely to be selected than the ones with lower fitness values (such as 11101). During *crossover*,

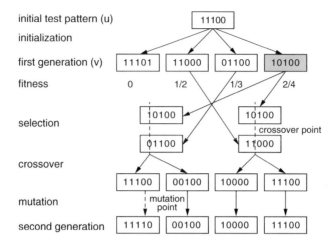

Fig. 9.3 Iterations of genetic algorithm for test vectors exploration [23]

Fig. 9.4 The comparison of cumulative distributions of sensitivities by GA versus MERS-s over 1000 Trojans [23]

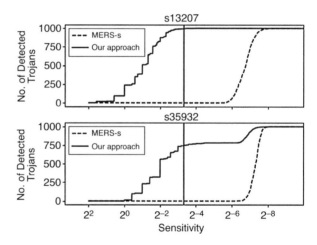

a single crossover point is randomly selected and crossover is performed on parents to produce two children. During *mutation*, a randomly selected position is mutated with a low mutation rate. For example, Fig. 9.3 shows only one mutation for four individuals.

Figure 9.4 shows the cumulative distribution of the sensitivities over 1000 Trojans in s13207 and s35932 for the proposed approach and MERS-s [13]. The x-axis is the sensitivity, y-axis is the number of Trojans that have sensitivities greater than x, and the vertical line represents 10% sensitivity. For example in s13207, almost all the Trojans have sensitivities greater than the sensitivity threshold in the approach, while in MERS-s this number is 0. In other words, if we assume the process variation is 10%, this approach can detect the majority of these randomly sampled Trojans with high confidence, while MERS-s cannot detect any of them.

9.4 Machine Learning Using Gate-Level Netlist Features

In [11, 27], the authors propose a static support vector machine (SVM) based hardware Trojan classification method for gate-level netlists. The proposed method classifies a set of the nets in a given netlist into Trojan nets and normal nets without using functional simulations. First, they extract the five hardware Trojan features, or Trojan features, based on the several known hardware Trojan-infected netlists. Then they apply machine learning to the extracted features. They consider the five Trojan features to be a five-dimensional vector and learn many five-dimensional vectors using a support vector machine (SVM). Finally, they can successfully classify a set of nets in a given unknown netlist into Trojan one and normal one by using the learned SVM classifier. The flow is shown in Fig. 9.5. Machine learning enables us to classify hardware Trojans automatically without simulating a given circuit or actually running it.

They considered the following five Trojan feature values for every target net n in a netlist to classify between Trojan nets and normal nets:

- LGF_i (Logic Gate Fan-ins): The number of inputs of the logic gates two-level away from the net n.
- FF_i (Flip-Flop Input): The number of logic levels to the nearest flip-flop input from the net n.
- FF_o (Flip-Flop Output): The number of logic levels to the nearest flip-flop output from the net n.
- PI (Primary Input): The minimum logic level from any primary input to the net n.
- PO (Primary Output): The minimum logic level to any primary output from the net n.

They used Gaussian radial basis function (RBF) for SVM learning. They decide the parameter values γ and C in SVM using learning data so that the true positive

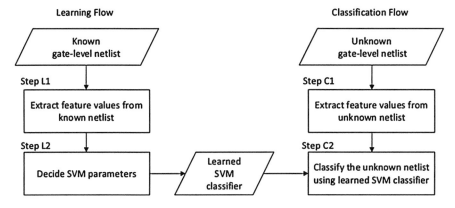

Fig. 9.5 The flowchart of learning and classification [11]

Table 9.1 Trojan detection on TRUST-HUB benchmarks [11]

Data name	Num. of all nets	Num. of Trojan nets	TPR	TNR
RS232-T1300	307	9	89%	26%
RS232-T1500	314	12	83%	24%
s35932-T200	6435	16	100%	59%
s38584-T100	7399	9	100%	62%
s38584-T300	9110	1730	89%	66%

rate (TPR) is maximized where TPR is defined by TPR = TP/(FN + TP). In this equation, TP shows the number of Trojan nets identified to be Trojan nets. FN shows the number of Trojan nets identified to be normal nets mistakenly. As shown in Table 9.1, the number of Trojan nets is relatively small compared to the total number of nets in a given netlist. This is because malicious third-party vendors tend to hide their presence in IC and try to pass the IC tests. Hence learning data for Trojan nets in SVM-based Trojan classification method may be much smaller than those for normal nets. It is very important to balance the learning data between Trojan nets and normal nets.

They balance the number of learned normal nets and the number of learned Trojan nets as follows: Assume that they have Nn normal nets and Nt Trojan nets. Then SVM has learned every normal nets once but every Trojan net (Nn/Nt) times. Overall, SVM has learned Nn normal nets and Nn Trojan nets because of weighting. As we can see from the Table 9.1, they can achieve good true positive rate, which means that their approach can detect more than 83% of all Trojan nets. However the true negative rate is very low, which means they can have more than 50% of normal nets mistakenly classified as Trojan nets. It would take huge amount of effort to deal with so many false positives. Their extended work [15] can achieve the average TPR (True Positive Rate) becomes 73%, and reduce the average FPR to 10%.

9.5 Trojan Detection Using Ring Oscillator Network

Recent work [18, 19] builds ring oscillator network (RON) on the power supply structure of an IC to detect hardware Trojan activity. In [19], principal component analysis (PCA) was used as a means of feature reduction. The data set contained the frequency data from eight ring oscillators (RO). A simple convex hull classification method was then used to classify each IC as either Trojan-free or into one of the 23 Trojan categories. While the RON is successful at detecting the difference between Trojan-free and Trojan-infected circuits, the FPR was nearly 50%. For these approaches to be practical in Trojan detection, we have to find ways to reduce the FPR.

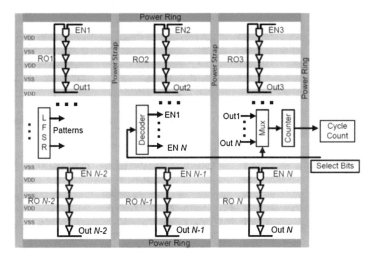

Fig. 9.6 The ring oscillator network (RON) structure [18]

As shown in Fig. 9.6, ROs consisting of inverters and a NAND gate for activation control are placed in a vertical orientation within the power structure of an IC. The ROs are then provided test patterns from a linear feedback shift register and a decoder. These outputs are then selected using a multiplexer and a counter registering the number of oscillations from the selected RO. The ROs frequency can then be derived from the number of oscillations. Any Trojan inserted into an IC will result in extra noise in the power supply structure that would not otherwise be present in a golden chip. By injecting the same test patterns into every IC, the Trojans should at least be partially active and thus cause extra noise. Since a ROs frequency is directly related to its power supply voltage, this Trojan caused power supply noise should propagate to the ROs frequency and result in differing measurements between clean and infected ICs. In [32], the authors can achieve above 90% classification accuracy, while reducing the false positive rate to below 10%.

The authors in [32] conducted experiments on eight FPGA boards (Nexys4 DDR development board). Each FPGA board is divided into four separate regions to increase the sample size. Each region is considered as an individual IC and Trojan, and the RON architecture is implemented in only a single portion at a time in order to make sure that one portion (or an individual IC) does not interfere another. A total of eight 41-stage ROs are used in each portion (i.e., IC). They distributed combinational and sequential Trojans in one portion randomly. The classifier was trained and optimized for three different sized data sets consisting of 6 chips, 12 chips, and 24 chips. Each sample size was then repeated for 20 trials and the average accuracy.

As can be seen in Table 9.2, the SVM is very accurate. However, when trained on fewer samples, it struggles with a high FPR. For 24 samples, it achieves a 97.4%

Table 9.2 SVM classification results

Metrics	Sample size		
	6 samples	12 samples	24 samples
TNR	0.445	0.605	0.929
FPR	0.555	0.355	0.071
FNR	0.017	0.023	0.023
TPR	0.983	0.977	0.977
Accuracy	0.940	0.946	0.974

Fig. 9.7 Statistical side-channel fingerprinting [21]

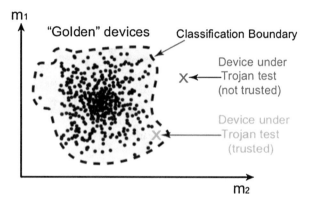

classification accuracy and a 7.1% FPR (Table 9.2) and outperform the results achieved in [18]. The authors believe that with a larger data set and increased training set sizes, the SVM can become more accurate and reduce the FPR even further. To further reduce the FPR, the authors proposed a simple voting ensemble by combining three classifiers: KNN + SVM + Naive Bayes. At the 24 chip training sample size, this can drive towards a 0% FPR, though it slightly reduces the accuracy.

9.6   Trojan Detection Using Side-Channel Fingerprints

For various hardware Trojan detection techniques in the literature, statistical side-channel fingerprints have been among the most heavily investigated ones. Starting with the global power consumption-based method presented in [1] and the path delay-based method introduced in [16], it became a popular direction to construct fingerprints (signatures) of ICs based on side-channel parameters and use these fingerprints to statistically assess whether an IC is contaminated by a hardware Trojan or not. This idea can be further applied to other side-channel signatures, such as power supply transient signals, leakage currents, temperature, wireless transmission power, as well as multi-parameter combinations [3].

Figure 9.7 illustrates how side-channel fingerprint works for hardware Trojan detection. A parametric signature of a chip is collected and compared to a trusted

Fig. 9.8 Golden-chip free side-channel fingerprinting [21]

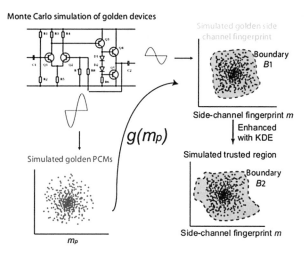

region in a multi-dimensional space. This trusted region is statistically established so that, despite the uncertainty incurred by process variations, the fingerprint of Trojan-free chips is expected to fall within this region while the fingerprint of Trojan-infested chips is expected to fall outside. Assuming availability of a representative set of trusted (i.e., golden) ICs, a classifier (e.g., neural network, support vector machine, etc.) can be trained to learn the boundary separating Trojan-free and Trojan-infested chips in the side-channel parametric space.

The drawback of these statistical side-channel fingerprints is the reliance on the availability of golden chips. In [21], the authors demonstrate that an almost equally effective trusted region can be learned through a combination of a trusted simulation model, measurements from process control monitors (PCMs) which are typically present either on die or on wafer kerf, and advanced statistical tail modeling techniques.

Figure 9.8 shows the PCM based fingerprints. Let $\overrightarrow{m_p} = m_{p,1}, \ldots, m_{p,n_p}$ denote the n_p-dimensional PCM measurement vector and $\overrightarrow{m} = m_1, \ldots, m_{n_m}$ denote the n_m-dimensional side-channel fingerprint vector, where n_p and n_m are the considered number of PCMs and side-channel fingerprints, respectively. Based on the n samples obtained by the Monte Carlo simulation, we can learn non-linear regression functions to map the PCM measurement pattern $\overrightarrow{m_p}$ to the values of each side-channel fingerprint of interest m_j. In other words, we train n_m regression functions $g_j : \overrightarrow{m_p} \rightarrow m_j$ as shown in Fig. 9.8. Such a simple and straightforward boundary has two major weaknesses: (1) Monte Carlo simulation produces few devices at the tails of the distribution, which is the area that matters the most when trying to establish a classification boundary, and (2) it has no anchor point in silicon and cannot reflect the process shifts that have taken place between the creation of the Spice simulation model and the current operating point of the foundry. In order to address the first weakness, the authors employ advanced tail modeling methods. For example, they add the kernel density function (KDE) $f(\overrightarrow{m})$ to generate an arbitrarily large volume of enhanced synthetic data. This can help accurately reflect the tails of the distribution $f(\overrightarrow{m})$. Based on the enhanced synthetic dataset of side-channel fingerprints, we can again use a one-class classifier to learn an improved classification boundary B2. Both B1 and B2 can be used for classification to identify Trojan-infected chips.

9.7   Runtime Trojan Detection in Multicore Processors

In this section, we discuss a case of machine learning for hardware Trojan detection in the routing network of a multicore processor. Assume that IP cores for processing cores and memories are secured, the communication network on many-core platform becomes an easy target of hardware Trojans. Attacks on a many-core router can affect network packet transfer rate and processing core availability, and may even cause interruption in core communication [5]. The router can be attacked externally through memory architecture interface, specialized core interface or internally by corrupting routing table. Typical attacks are traffic diversions, routing loops, core spoofing attacks. All three attacks are denial of service (DoS) attack, wherein a specific core under attack is made unavailable.

- *Traffic Diversion Attack:* Under this attack, the router selects a random core to transfer the packet. This attack affects the deadline for other cores, which are dependent on the packet under attack.
- *Routing loop attack:* Under this attack, the packets are routed back to the source core. The source core is made unavailable to other communicating cores, thereby causing latency in other core transfers.
- *Core Spoofing Attack:* This attack transfers all packets to randomly chosen destination. The attack could saturate the core and make it unavailable to other cores.

In [20], the authors propose a runtime Trojan detection architecture for a custom many-core based on machine learning technique. Collecting relevant data based on hardware behavior analysis is the first step for machine learning to succeed. Relevant feature selection will increase the accuracy of Trojan detection and aid the hardware implementation as well. Removing irrelevant features will reduce dataset thereby decreasing hardware complexity and area overhead. Therefore, we select relevant features based on feature correlation analysis. The authors consider the following features for Trojan detection:

- Source Core: Source core number.
- Destination Core: Destination core number.
- Packet Transfer Path: Packet transfer between the two cores has a unique path which alters in case of Trojan. For a 88 NoC, i.e., many-core architecture with 64 processing cores, it has three levels of router hops. The highest number of hops to be traveled by packet can be 5 for inter-cluster communication.
- Distance: At each router hop, distance vector is incremented by 1. For example, when core 11 is transferring packet to core 62, distance vector is incremented at $R0\_2$, $R1\_0$, $R2\_0$, $R1\_3$, $R0\_3$. Since there will be six vertices (five router hops and one processing core) and five edges, the distance is 6.

Figure 9.9 shows the 64-core test setup implemented on Xilinx Virtex-7 FPGA. The setup consists of three major modules: (1) 64-core many-core architecture, (2) attack detection module, (3) Trojan insertion module. The authors used a bio-medical seizure detection algorithm as the application for testing out their Trojan detection approach. The seizure detection algorithm is mapped on the 64-core architecture. It takes 458 inter-cluster and 1088 intra-cluster communications, where

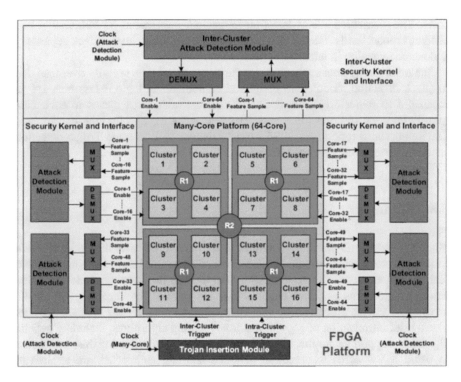

Fig. 9.9 Test setup for many-core platform with 64 processing cores on Xilinx Virtex-7 FPGA [20]. Each cluster has a Level 0 router $R0$ to connect 4 cores, and there are 4 Level 1 routers $R1$ to connect all 16 $R0$ routers, and finally the Level 2 router $R2$ connects all 4 $R1$ routers

Table 9.3 Area overhead on Xilinx Virtex-7 FPGA

Logic utilization	Many-core only	Many-core with SVM	Security overhead
Slice count	55,072	55,220	148 (0.26%)
Register count	49,472	49,830	358 (0.72%)
LUT count	142,008	142,281	273 (0.2%)
Memory count	11,244	11,244	–

each cluster consists of four processing cores to execute the application. For each core-to-core communication, the packet is generated with data that needs to be transferred and the address of the destination core. *FeatureSample* is updated at each communication hop to build the feature vector for classification. At the source core, *FeatureSample* is updated with two features, i.e., source core, destination core, and other feature attributes are initialized as zeros. At each router hop, *FeatureSample* updates other attributes, i.e., path and distance. Finally at destination router before the destination core, *FeatureSample* is transferred to the attack detection module which is designed based on SVM. Many-core architecture is fully placed and routed in Xilinx Virtex-7 FPGA, which consists of 64 processing cores. The distributed attack detection framework implements SVM kernel at two different router levels. Each SVM kernel will detect intra-cluster attack separately and hence reduction in latency of operations.

Table 9.3 shows area analysis and security overhead. The security overhead due to attack detection module and peripheral combinational logic is less than 1% as compared to router architecture. Security kernel adds three cycles to each data transfer between 16-cores (intra-cluster), whereas three cycles for inter-cluster data. The authors also report that SVM kernel achieves 93% average Trojan detection accuracy for randomly inserted Trojans.

9.8   Summary

Machine learning algorithms are natural fit for hardware Trojan detection, wherein Trojan affected designs should be distinguished from good designs. When suitable features are extracted and good dataset feed to train the learning model, machine learning algorithms can find patterns that are beyond the capability of human analysis of circuit properties or side-channel signatures. In this chapter, we first give a brief introduction of popular machine learning algorithms, including genetic algorithm, support vector machines, liner regression, etc. Following this, we discuss in detail a few cases where these machine learning approaches are applied to detect hardware Trojans. We discussed five different scenarios in applying machine learning algorithms for Trojan detection: (1) test vector generation using genetic algorithm, (2) machine learning approach using circuit features, (3) utilization of ring oscillator network, (4) Trojan detection using side-channel fingerprints,

(5) a runtime Trojan detection in routing network of multicore processors. There are many more possibilities of applying machine learning for detecting hardware security vulnerabilities.

References

1. D. Agrawal, S. Baktir, D. Karakoyunlu, P. Rohatgi, B. Sunar, Trojan detection using IC fingerprinting, in *IEEE Symposium on Security and Privacy* (2007), pp. 296310
2. M.E. Amyeen et al., Evaluation of the quality of N-detect scan ATPG patterns on a processor, in *International Test Conference* (2004)
3. S. Bhunia et al., Hardware Trojan attacks: threat analysis and countermeasures, in *IEEE Special Issue on Trustworthy Hardware* (2014)
4. R. Chakraborty et al., MERO: a statistical approach for hardware Trojan detection, in *CHES Workshop* (2009)
5. S. Charles, Y. Lyu, P. Mishra, Real-time detection and localization of DoS attacks in NoC based SoCs, in *Design Automation and Test in Europe (DATE)* (2019)
6. J. Cruz et al., Hardware Trojan detection using ATPG and model checking, in *International Conference on VLSI Design* (2018)
7. J. Cruz, Y. Huang, P. Mishra, S. Bhunia, An automated configurable Trojan insertion framework for dynamic trust benchmarks, in *Design Automation and Test in Europe (DATE)* (2018)
8. J. Cruz, P. Mishra, S. Bhunia, The metric matters: how to measure trust, in *Design Automation Conference (DAC)* (2019)
9. F. Farahmandi, Y. Huang, P. Mishra, Trojan localization using symbolic algebra, in *Asia and South Pacific Design Automation Conference (ASP-DAC)* (2017), pp. 591–597
10. X. Guo, R.G. Dutta, P. Mishra, Y. Jin, Automatic code converter enhanced PCH framework for SoC trust verification. IEEE Trans. Very Large Scale Integr. VLSI Syst. **25**(12), 3390–3400 (2017)
11. K. Hasegawa, M. Oya, M. Yanagisawa, N. Togawa, Hardware Trojans classification for gate-level netlists based on machine learning, in *IEEE 22nd International Symposium on On-Line Testing and Robust System Design (IOLTS)*, Sant Feliu de Guixols (2016), pp. 203–206
12. K. Hasegawa, Y. Shi, N. Togawa, Hardware Trojan detection utilizing machine learning approaches, in *Proceedings - 17th IEEE International Conference on Trust, Security and Privacy in Computing and Communications and 12th IEEE International Conference on Big Data Science and Engineering, Trustcom/BigDataSE 2018* (2018), pp. 1891–1896
13. Y. Huang, S. Bhunia, P. Mishra, MERS: statistical test generation for side-channel analysis based Trojan detection, in *ACM Conference on Computer and Communications Security* (2016)
14. Y. Huang, S. Bhunia, P. Mishra, Scalable test generation for Trojan detection using side channel analysis, in *IEEE Transactions on Information Forensics and Security* **13**(11), 2746–2760 (Nov. 2018)
15. T. Inoue, K. Hasegawa, Y. Kobayashi, M. Yanagisawa, N. Togawa, Designing subspecies of hardware Trojans and their detection using neural network approach, in *IEEE 8th International Conference on Consumer Electronics - Berlin, ICCE-Berlin* (2018)
16. Y. Jin, Y. Makris, Hardware Trojan detection using path delay fingerprint, in *IEEE International Workshop on Hardware-Oriented Security and Trust* (2008), pp. 5157
17. Y. Jin, Y. Makris, Hardware Trojans in wireless cryptographic ICs, in *IEEE Design and Test of Computers*, **27**(1), 2635 (2010)

18. N. Karimian, F. Tehranipoor, D. Forte, Md.T. Rahman, Genetic algorithm for hardware Trojan detection with Ring Oscillator Network (RON), in *IEEE International Conference on Technologies for Homeland Security* (2015)
19. S. Kelly, X. Zhang, M. Tehranipoor, A. Ferraiuolo, Detecting hardware Trojans using on-chip sensors in an ASIC design. J. Electron. Test. **31**(1), 11–26 (2015)
20. A. Kulkarni, Y. Pino, T. Mohsenin, SVM-based real-time hardware Trojan detection for many-core platform, in *17th International Symposium on Quality Electronic Design (ISQED)*, March (2016)
21. Y. Liu, K. Huang, Y. Makris, Hardware Trojan detection through golden chip-free statistical side-channel fingerprinting, in *Proceedings of the 51st Annual Design Automation Conference (DAC '14)*. ACM, New York, Article 155 (2014), p. 6
22. Y. Lyu, P. Mishra, A survey of side channel attacks on caches and countermeasures. J. Hardw. Syst. Secur. **2**, 33–50 (2018)
23. Y. Lyu, P. Mishra, Efficient test generation for Trojan detection using side channel analysis, in *Design Automation and Test in Europe (DATE)*, Florence, Italy, March 25–29 (2019)
24. M. Mitchell, *An Introduction to Genetic Algorithms* (MIT Press, Cambridge, 1996)
25. M. Mohri, A. Rostamizadeh, A. Talwalkar, *Foundations of Machine Learning* (The MIT Press, Cambridge, 2012). ISBN 9780262018258
26. S. Narasimhan, D. Du, R. Chakraborty, S. Paul, F. Wolff, C. Papachristou, K. Roy, S. Bhunia, Multiple-parameter side-channel analysis: a non-invasive hardware Trojan detection approach, in *IEEE International Symposium on Hardware-Oriented Security and Trust* (2010), pp. 1318
27. M. Oya, Y. Shi, M. Yanagisawa, N. Togawa, A score-based classification method for identifying hardware-Trojans at gate-level netlists, in *Proceedings of the 2015 Design, Automation and Test in Europe (DATE '15)*. EDA Consortium, San Jose (2015), pp. 465–470
28. E.M. Rudnick et al., A genetic algorithm framework for test generation. IEEE Trans. Comput. Aided Des. Integr. Circuits Syst. **16**, 1034–1044 (1997)
29. S. Saha et al., Improved test pattern generation for hardware Trojan detection using genetic algorithm and boolean satisfiability, in *Cryptographic Hardware and Embedded Systems – CHES* (2015)
30. A. Waksman, M. Suozzo, S. Sethumadhavan, Fanci: identification of stealthy malicious logic using boolean functional analysis, in *ACM SIGSAC Conference on Computer & Communications Security* (2013), pp. 697–708
31. F. Wolff et al., Towards Trojan-free trusted ICs: problem analysis and detection scheme, in *Design, Automation and Test in Europe* (2008)
32. K. Worley, Md. T. Rahman, Supervised machine learning techniques for Trojan detection with ring oscillator network (2019). Available at arXiv:1903.04677v1

Chapter 10
Trojan Detection Using Dynamic Current Analysis

10.1 Introduction

Existing test generation solutions for hardware Trojan detection [18] can be broadly classified into two categories: (1) logic testing and (2) side-channel analysis. In logic testing approach, directed tests are generated to activate rare events in a circuit and propagate the malicious effect of a Trojan in logic values to observable outputs. Such approaches are known to be more effective in detecting ultra-small Trojans (typically a few gates in size). The main challenge with logic testing approaches, however, is the difficulty to trigger a Trojan and observe its effect, particularly in the presence of complex sequential Trojans, and the inordinately large number of possible Trojan instances that an adversary can exploit. On the other hand, side-channel analysis approaches depend on the measurement and analysis of physical "side-channel" parameters like power signature or path delay of an IC in order to identify a structural change in the design. Unlike logic testing, these approaches do not require Trojan activation in order to detect them. Side-channel analysis (SCA), primarily based on supply current, has been extensively investigated by large number of research groups and various solutions to increase the signal-to-noise (SNR) have been proposed. A disadvantage of SCA arises from the large process variations (e.g., $20\times$ leakage power and 30% delay variations in 180 nm technology [3]) which can potentially mask the minute effect of a Trojan in the measured side-channel parameter. A solution to the sensitivity problem can be achieved by judicious test generation approach that aims at maximizing the sensitivity for an arbitrary Trojan in unknown circuit location. In this chapter, we focus on transient current or power as side-channel parameter of interest. Some of the concepts however can be applied to other side-channel parameters. To maximize sensitivity of a given Trojan, one needs to amplify activity inside the Trojan circuit and simultaneously minimize the background activity (i.e., activity in the original circuit). We present a novel statistical test generation framework that can maximize the detection sensitivity for an arbitrary Trojan.

© Springer Nature Switzerland AG 2020

F. Farahmandi et al., *System-on-Chip Security*,
https://doi.org/10.1007/978-3-030-30596-3_10

The rest of the chapter is organized as follows. Section 10.2 presents related work in side-channel analysis and functional test generation for Trojan detection. Section 10.3 describes the proposed MERS test generation algorithm and the test reordering algorithms to improve sensitivity of side-channel analysis. Section 10.4 describes the experimental setup and presents results on a set of ISCAS benchmarks with detailed analysis. Section 10.5 presents results for two large designs (AES cipher and DLX processor). Section 10.6 concludes the chapter.

10.2   Related Work

The underlying assumption for Trojan insertion is that an adversary is fully aware of the design functionality and therefore can hide the Trojan in a hard-to-find place. One way to address this issue is to obfuscate [4] or encrypt [9] the design such that the adversary cannot figure out the actual functionality and therefore cannot insert the Trojan in a covert manner. Unfortunately, smart attacker can effectively bypass both obfuscation [23] and encryption [27] methods. A promising direction is to develop efficient techniques for hardware Trojan detection. Prior research on Trojan detection can be classified into two broad categories: side-channel analysis and functional test generation.

Side-channel analysis approaches [8] are based on analysis of side-channel signatures such as circuit transient current [1, 2, 25], power consumption [22, 29], path delay [14], or intermediate values from debug infrastructure [11]. The basic idea is to compare the side-channel signature with the pre-characterized golden value for a Trojan-free IC (or a model of the IC). If the observed value of the measured parameter differs by more than a threshold from the golden value, the presence of a Trojan is suspected. Unfortunately, side-channel analysis has a common issue, i.e., the sensitivity of side-channel signatures is susceptible to thermal and process variations. Therefore, it would be difficult to detect small combinational Trojans. We also rely on transient current (switching activity) to identify Trojans.

Compared with [1, 2, 25], the proposed approach can greatly increase the side-channel sensitivity of Trojan of any type or size, because we take advantage of functional testing. In other words, the test vectors are generated in a statistical way, and they are more effective in creating switching in Trojan, as well as reducing background switching. The approach proposed by Banga and Hsiao [1] partitions a design into circular regions (with a center and radius) for side-channel analysis. A region is a group of flip-flops along with combinational gates connecting them. However, there are two major drawbacks with their partitioning approach. First, there are thousands of regions identified even for a small ISCAS89 benchmark s3271. It may be infeasible to generate targeted tests for each of the regions in large designs. Next, regions identified by their approach may overlap with each other, while the proposed approach can ensure the regions are disjoint. Banga et al. proposed in [2] to partition a circuit into flip-flop groups based on structural connectivity. However, the scalability of the approach to large designs with datapaths and control structure is limited. Moreover, it is difficult to judge their effectiveness since they are only tested on very small circuits. Salmani et al. [25] proposed a layout-aware approach for improving localized switching to detect Trojan. Their approach is based on reordering the scan cells (flip-flops) in the chip, which is orthogonal to the proposed approach of test generation for improving switching.

Another category of Trojan detection approaches is to generate functional test patterns that are likely to fully activate the Trojans [7, 12, 13, 16, 17]. These approaches can overcome the effect of thermal and process variations on side-channel signals. They rely on the fact that an adversary will choose a trigger condition for the Trojan using a set of rare nodes. Various approaches tried to maximize the rare node activation to increase the likelihood of activating Trojans. ATPG for Trojan detection is investigated in [6, 30]. A major problem with ATPG based Trojan detection methods is the scalability issue. ATPG can be used to activate a Trojan if all the triggers are known. However, this is not feasible for Trojan detection since Trojans are likely to have unknown number of triggers hidden at stealthy locations. It would be practically infeasible to use ATPG to test all possible trigger conditions. MERO [5] takes the advantage of N-detect test [21] to maximize the trigger coverage by activating the rare nodes. The test generation ensures that each of the nodes gets activated to their rare values for at least N times. It is shown that if N is sufficiently large, a Trojan with trigger condition based on these rare nodes is likely to be activated by the generated testset. Saha et al. [24] improve the test pattern generation of MERO [5] by using genetic algorithm and Boolean satisfiability for ATPG. Their approach could more effectively propagate the payload of possible Trojan candidates. A design-for-test (DFT) infrastructure technique by Salmani et al. [26] inserts dummy flip-flops to increase the transition probability of low-transition nets, and therefore increases the side-channel sensitivity for Trojan detection. Zhou et al. [32] further improved their approach by selecting the most beneficial nets to insert dummy flip-flops based on fanout analysis. Farahmandi et al. [10] attempted to localize Trojan using symbolic algebra from a formal verification approach, while it is not scalable to large circuits.

Direct application of test generation approaches is not suitable for improving side-channel sensitivity for Trojan detection. The objective of increasing side-channel sensitivity is very different from the ones in both MERO [5] as well as its enhanced version by Saha et al. [24]. Unlike these existing techniques, the proposed approach requires the creation of a pair of test vectors to maximize switching in rare nodes. This algorithm creates multiple excitation of rare switching which is important in making side-channel based Trojan detection effective. The initial idea [12] does not provide a scalable test generation framework for different DFT structures. Moreover, it is important to simultaneously minimize the background switching to maximize the relative switching.

The proposed test generation method also originates from N-detect test. Compared with MERO [5], which focuses on logic testing with N-detect test, we target generating vectors for side-channel analysis. The primary difference is that MERO tries to assign rare values (0 or 1), whereas the proposed approach tries to assign rare transitions (0 → 1, or 1 → 0). Specifically, they have three important differences. First, MERO's goal is to generate tests which can fully trigger the Trojan and observe the propagated Trojan effect. This algorithm aims at creating more switching in possible Trojan triggers to greatly improve the side-channel sensitivity and expose hideous Trojans. Next, MERO's approach is mostly limited to combinational Trojans with smaller number of triggers. MERO is not effective for sequential Trojans or larger Trojans. The proposed approach focuses on switching of rare nodes, which makes it effective to any type/size of Trojans hidden at any location. Finally, by utilizing functional and structural partitioning, the proposed approach is scalable to large designs with a large number of rare nodes or possible triggering conditions.

10.3   Test Generation for Side-Channel Aware Trojan Detection

In this section, we present the proposed methodology for side-channel aware test generation in detail. The methodology is based on the concept of statistically maximizing the switching activity in all the rarely triggered circuit nodes. The effectiveness of a test pattern for side-channel analysis is measured in two ways: (1) the ability to create most switching inside a Trojan or to activate a Trojan; (2) the ability to create high Trojan-to-circuit switching. We measure *DeltaSwitch* as the switching introduced by the Trojan, which is the difference of number of switches between the golden circuit and the Trojan-infected circuit. We measure *RelativeSwitch* as the ratio of *DeltaSwitch* to the total number of switches (*TotalSwitch*) in the golden circuit. An effective test vector should be capable of creating large *DeltaSwitch*, and more importantly it should create large *RelativeSwitch*, as it is directly related to the sensitivity for side-channel analysis.

$$RelativeSwitch = DeltaSwitch/TotalSwitch \qquad (10.1)$$

As shown in Fig. 10.1, we provide the overview of the workflow for scalable test generation for side-channel aware Trojan detection. We first simulate the circuit to get the rare nodes, which have low probability to be 0/1. We partition the design into regions, apply the MERS approach to generate tests, and also reorder the tests for each region. The vectors from all regions are combined into a test suite that can create high relative switching for arbitrary Trojans in the design.

Rare Nodes Identification In the experiments, we simulate the circuit with 100,000 random vectors and note down the probability of values for internal nodes. Nodes with probability less than the rare threshold are identified as rare nodes. These rare nodes are the candidates for Trojan triggers. We sample stealthy Trojans with triggers from the rare nodes for evaluation of the test generation approach.

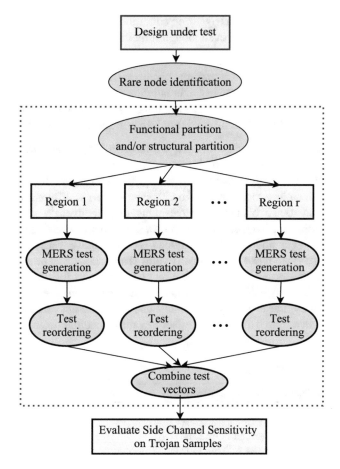

Fig. 10.1 Scalable test generation for side-channel analysis based Trojan detection

Design Partitioning A major challenge of large designs is that the supply current of a golden chip for a high-activity vector can be very large compared to the additional current consumed by a small Trojan. If we can carefully partition a circuit into nearly isolated regions (i.e., with low connectivity between them), we can more effectively generate tests for each region. After partitioning the design, test generation can target on the rare nodes inside each region, but also try to avoid creating too many background switching.

Test Generation The test generation approach (MERS) is based on creating a set of test vectors for each candidate rare node individually to have *rare switching* multiple (at least N) times. MERS utilizes the principle of N-detect [21] tests to increase the likelihood of partially or fully activating a Trojan. MERS can generate a high-quality testset for these rare nodes individually to have rare switching for N times. If N is sufficiently large, a Trojan with triggering conditions from these rare nodes is likely to have high switching activity even though it might not be fully activated.

Test Reordering   The order of test vectors matters as we are counting the switching between two vectors. The goal is to further improve the side-channel sensitivity. The challenge is to keep the high-quality in creating switching on rare nodes, and at the same time to reduce the background switching. We introduce Hamming-distance based reordering and simulation-based reordering to resolve this challenge.

Testset Evaluation   In the experiments, we insert a Trojan into the design, then apply all test vectors in the combined testset. The side-channel sensitivity is reported as the maximum relative switching of the testset. To show that the proposed approach has good coverage on Trojans hidden at different locations, we experimented on 1000 Trojan samples to evaluate the effectiveness of testsets. From a pool of potential rare Trigger nodes, a Trojan of given size is created by randomly choosing the trigger nodes and the payload. After that we verify if this trigger condition and payload make a functionally valid Trojan, i.e., it can be activated using a valid input condition and its malicious effect propagates to any observable output. Thus, we consider only valid random Trojans in the evaluation. The statistical nature of MERS ensures that even if an adversary chooses different locations or trigger conditions for inserting Trojans, the testset can maximize the detection sensitivity for them.

Algorithm 18: Scalable test generation

    Input: Circuit under test
    Output: Test patterns for Trojan Detection
1 // Rare nodes identification
2 Simulation to identify nodes with low probability
3 Generate Trojan samples with triggers from rare nodes
4 // Design Partitioning
5 **if** *Design partition is enabled* **then**
6   **if** *Design has natural sub-modules* **then**
7 | Functional partition into regions.
8   **end**
9   **if** *Design (or any region) is large* **then**
10 | Structural partition based on connectivity.
11   **end**
12 **end**
13 // Test Generation
14 **for** *each region* **do**
15   Test generation with MERS (Algorithm 19)
16   Test reordering with Algorithm 20 or 21
17 **end**
18 // Evaluation of Test Patterns
19 Combine the testsets from all regions
20 Evaluate side-channel sensitivity on Trojan samples

Algorithm 18 shows the steps for scalable test generation on large designs. We first simulate the circuit to identify rare nodes and generate stealthy Trojan samples. If design partitioning is enabled, the design is partitioned according to natural boundaries based on functionality. If the design has no such natural boundaries, or the partitioned region is still too large, we can perform structural partitioning based on circuit connectivity. For each region, we apply high-quality test generation approach MERS (Algorithm 19), followed by test reordering (Algorithm 20 or 21) for further improvement in side-channel sensitivity. Finally, the test patterns from all region will be combined together to evaluate the effectiveness of the proposed test generation approach on the Trojan samples.

10.3.1   Design Partitioning

There are at least three advantages of dividing a large design into smaller regions. (1) For a designated region, region-based MERS (Algorithm 19) will only target the rare nodes in that region to have rare switching for N times. The quality of tests is likely to improve and many rare nodes can achieve rare switching a lot more than N times. (2) The rare nodes outside of the designated region will be ignored. Since the test generation process does not try to switch those rare nodes, it is likely to create fewer switching in the outside regions and reduce the background switching. (3) Assuming that the sequential circuits are equipped with scan-chains, we can shift 0's into the flip-flops (the pseudo primary inputs) that are outside of the designated region. This can further reduce the background switching of other regions.

The partitioning approach should divide the design into regions, which have minimum inter-connections between them. In other words, we want each region to be functionally independent or have as few connections as possible to other regions, so that the test generation process can increase the activity of one region (or few regions) while minimizing the activity of all others. A complex circuit under test usually comprises several functional modules (or regions), which are interconnected according to their input/output dependencies. For the example in Fig. 10.2a, the DLX processor has four pipeline stages (IF, ID, EXE, and MEM). It can be naturally partitioned into four regions according to the functional modules: Fetch, Decode, Execute, and Memory. We can fill the pipeline such that the different pipeline stages are activated one at a time during test generation. An alternative to functional partitioning is structural partitioning as shown in Fig. 10.2b. Structural partitioning is the only choice when functional partitioning is not possible (e.g., flattened netlist). Structural partitioning can use hypergraph partitioning approach [15] or any other region-based partitioning approaches [1]. It is important to note that this approach can effectively combine both partitioning techniques. For example, after functional partitioning has been performed on DLX processor (Fig. 10.2a), structural partitioning (Fig. 10.2b) can be applied on the Decode module (accounting for 71% of the whole design area) to further partition it into smaller regions.

Fig. 10.2 (**a**) Functional
partitioning: a design with
functional modules can be
naturally partitioned. (**b**)
Structural partitioning: a flat
design can be structurally
partitioned to find regions
with minimum
inter-connections being cut

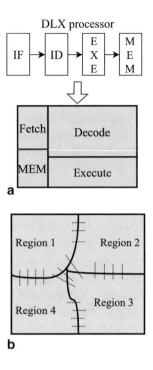

We use structural partitioning to improve the side-channel sensitivity for the three ISCAS sequential benchmarks: s13207, s15850, and s35932. Since these benchmarks are flat netlists (i.e., we cannot easily identify any functional regions), we use hypergraph partitioning [15] to find relatively isolated regions. The circuit is transformed into a hyper-graph, where each gate is a vertex and the set of gates sharing an edge is a hyper-edge. The goal of partitioning is to divide the graph into two partitions of roughly same size[1] (each contains 45–55% of the total number of vertices). The constraint is that the minimal number of hyper-edges will be *cut* by the partition process. This constraint is to ensure that the vertices inside each region have high connectivity, while the connectivity between regions is minimal. This is a well-studied problem in hypergraph, and we used the tool from [15] to satisfy this purpose. The whole design is first partitioned into two regions of almost equal size with minimal hyper-edge cuts. Each of the two regions can be partitioned into two smaller sub-regions, and so on. In other words, we can partition the design in a recursive manner to have two regions, four regions, and eight regions.

---

[1] We have performed structural partitioning [15] with partition factor $p = 50\%$, which generally achieves better SCS than a skewed partition. If the designer can afford the cost, different partition factor p can be explored.

Algorithm 19: Multiple excitation of rare switching (MERS)

Input: Circuit netlist (targeted region), rare switching requirement (N), the list of rare nodes
   ($R = \{r_1, r_2, \ldots, r_m\}$), the list of random patterns ($V = \{v_1, v_2, \ldots, v_n\}$)
Output: MERS test patterns (T)

for *each random vector v in V* **do**
   Simulate the circuit with the input vector v
   Count the number of nodes (R_V) in R with their rare values satisfied
end
Sort vectors in V in descending order of R_V
for *each node r_i in R* **do**
   Set its rare switching counter (S_i) to 0
end

Initialize previous vector t_p as a vector of all 0's
for *each vector v_j in V* **do**
   Simulate the circuit with vector pair (t_p, v_j)
   Count the number of rare switches (R_S)
   Set $v'_j = v_j$
   for *each bit in v'_j* **do**
      Mutate the bit and re-simulate the circuit with vector pair (t_p, v'_j)
      Count the number of rare switches (R'_S)
      if $R'_S > R_S$ **then**
         Accept the mutation to v'_j
      end
   end
   Update S_i for all nodes in R due to vector v'_j
   if v'_j *increases S_i for at least one rare node* **then**
      Add the mutated vector v'_j to T
      Set $t_p = v'_j$
   end
   if $S_i \geq N$ *for all nodes in R* **then**
      Break
   end
end
return *MERS test patterns T*

10.3.2   Multiple Excitation of Rare Switching (MERS)

The basic idea of MERS is that if we can make a rare node switch N times where N is sufficiently large, it significantly improves the chance of switching in a Trojan associated with that rare node. The **rare switching** in the algorithm specially refers to a rare node switching from its non-rare value to its rare value. The reason to choose this criteria is twofold: (1) it is more difficult to switch from non-rare to rare value than from rare to non-rare value; (2) it defines the switching between the previous vector and the current vector, and it usually helps to create an extra switching between the current vector and the next vector. This will increase the probability of switching of a Trojan which has rare nodes as its trigger conditions.

This approach is also applicable to sequential Trojans, which requires the rare condition to occur a certain number of times to be fully triggered.

Algorithm 19 shows the steps of MERS to generate high-quality tests for creating switching in rare nodes, so as to assist side-channel analysis for hardware Trojan detection. The algorithm is fed with the golden circuit netlist, the list of random test patterns (V) and a list of rare nodes (R) (which is obtained by random vector based circuit simulation beforehand). First, we simulate each random pattern and count the number of rare nodes (R_V) that take their rare values. We sort the random patterns in descending order of R_V, which means that the vector with ability to activate the most number of rare nodes goes first. Next, we initialize the rare switching counter S_i for each rare node to 0. In the next step, we mutate vectors from the random pattern set to generate high-quality tests. We mutate the current vector one bit at a time and we accept the mutated bit only if the mutated vector can increase the number of nodes to have rare switching. In this step, only those rare nodes with $R_S < N$ are considered. The mutation process repeats until each rare node has achieved at least N rare switches. The output of the test generation process is a compact set that improves the switching capability in rare nodes, compared to random patterns. The complexity of the algorithm is $O(n * m)$, where n is the total number of test vectors mutated during the process, and m is the number of bits in primary inputs. The runtime to generate MERS tests can be found in Table 10.1.

The testset generated by MERS is expected to be very effective in increasing the likelihood of rare nodes to switch and thus increasing the activities in Trojans. In other words, MERS testset is capable of maximizing the DeltaSwitch (the numerator in Eq. 10.1). Further extension of this work also explores using genetic algorithm to mutate the vector pairs, which can more thoroughly search the test vector space. Readers can refer to [17].

MERS testset is already a very high-quality testset in terms of criteria for DeltaSwitch. However, MERS testset also creates more switching in other parts of the circuit, when it is making efforts to switch rare nodes. This characteristic of increased TotalSwitch would be further illustrated in the Sect. 10.4. In order to

Table 10.1 Runtime comparison for MERO [5], MERS-h and MERS-s, with $N = 1000$, rarethreshold $= 0.1$

Benchmark	Nodes (rare/total)	Runtime (s)		
		MERO [5]	MERS-h	MERS-s
c2670	63/1010	30051.53	13378.1	18296.09
c3540	331/1184	9403.11	6106.94	24264.45
c5315	255/2485	80241.52	45607.01	84669.78
c6288	45/2448	15716.42	4154.93	6957.47
c7552	306/3720	160783.37	81431.09	144908.08
s13207	592/2504	23432.04	12876.97	41576.67
s15850	679/3004	39689.63	20631.58	58084.93
s35932	896/6500	29810.49	7335.27	38496.78
Average	396/2857	48,641	23,940	52,157

maximize relative switching, we need to have TotalSwitch in control as well. In the following subsections, we propose two methods to tune the MERS testset, so that it can: (1) still be effective for DeltaSwitch, (2) reduce TotalSwitch and improve the effectiveness for RelativeSwitch. The first method is a heuristic approach based on hamming distance of test vectors, which can reduce the total switching. The second one is simulation based, in which we try to balance the rare switching and the total switching while we explore all the candidate vectors.

10.3.3   Test Reordering

10.3.3.1   Hamming Distance Based Reordering

If two consecutive input vectors have the same values in most bits, it is very possible that the internal nodes will also have a lot of values in common. A simple heuristic to reduce total switching in circuit is to have similar input vectors. We use the Hamming distance between two vectors to represent the similarity. Algorithm 20 shows an approach to reorder the testset by Hamming distance. The algorithm is a greedy approach to explore all candidate vectors and take the best one in terms of Hamming distance. We first check the Hamming distances between the previous vector and all the remaining vectors, then we select the vector which has the minimum Hamming distance as the next vector. The time complexity of Algorithm 20 is $O(n^2)$, where n is the testset size. Fortunately, it is of low cost to calculate the Hamming distance between two input vectors, so the actual runtime is very short.

Algorithm 20: Tests reordering by hamming distance (MERS-h)

Input: List of Test Patterns ($T_{orig} = \{t_1, t_2, \ldots, t_n\}$) produced by Algorithm 1
Output: Improved Test Patterns (T_{hamm})

Initialize $T_{hamm} = \{\}$
Initialize previous test t_p as a vector of all 0's
while T_{orig} *is not empty* **do**
    $min_{dist} = int\_max$
    $best_{idx} = -1$
    for **all** *remaining tests* t_j *in* T_{orig} **do**
        if $min_{dist} > hamming\_dist(t_p, t_j)$ **then**
            $min_{dist} = hamming\_dist(t_p, t_j)$
            $best_{idx} = j$
        end
    end
    Add $t_{best_{idx}}$ *to the end of* T_{hamm}
    Remove $t_{best_{idx}}$ *from* T_{orig}
    Update $t_p = t_{best_{idx}}$
end
return T_{hamm}

10.3.3.2   Simulation-Based Reordering

The reordering problem to improve the relative switching is actually a multi-objective optimization problem: maximize the $DeltaSwitch$ and minimize the $TotalSwitch$ as in Eq. 10.1. We do not know the $DeltaSwitch$, because the location and type of the Trojan is unknown. However, rare switching between two vectors is a good indicator for $DeltaSwitch$, which means a large number of rare switching would imply a large $DeltaSwitch$ in Trojan. We redefine the optimization goal as to maximize the rare switching and minimize the total switching at the same time between vector pairs. We formalize the problem as shown in Eq. 10.2. We need to explore the best weights to balance between the two objectives:

$$maximize \quad (w_1 * RareSwitch - w_2 * TotalSwitch) \tag{10.2}$$

We propose an approach as shown in Algorithm 21 based on real simulation of the test vectors to maximize the combined objective. We introduce a concept of $profit$ to indicate the fitness of a test vector to follow the previous test vector. $profit$ is defined as $(C * RareSwitch - TotalSwitch)$, where C is the ratio of two weights w_1 and w_2. It is meant to maximize the rare switching (activity in Trojan circuits) and minimize the total switching of the whole circuit. In the experiment section, we will explore different weight ratios and check the influence of weight ratios on side-channel sensitivity.

Algorithm 21: Tests reordering by simulation (MERS-s)

Input: List of Test Patterns ($T_{orig} = \{t_1, t_2, \ldots, t_n\}$) produced by Algorithm 1
Output: Improved Test Patterns (T_{sim})

Initialize $T_{sim} = \{\}$
Initialize previous test t_p as a vector of all 0's
while T_{orig} *is not empty* **do**
    $max_p = int\_min$
    $best_{idx} = -1$
    for all *remaining tests* t_j *in* T_{orig} **do**
        Simulate the circuit with vector pair (t_p, t_j)
        Count the number of RareSwitch and TotalSwitch
        $profit = C * RareSwitch - TotalSwitch$
        if $max_p < profit$ **then**
            $max_p = profit$
            $best_{idx} = j$
        end
    end
    Add $t_{best_{idx}}$ to the end of T_{sim}
    Remove $t_{best_{idx}}$ from T_{orig}
    Update $t_p = t_{best_{idx}}$
end
return T_{sim}

Algorithm 21 shows the proposed approach to tune the testset by simulation with *profit* as a reordering criterion. By exhaustively checking the *profit* between the previous vector and all the remaining vectors, we select the vector which has the maximum *profit* as the next following vector. The time complexity of Algorithm 21 is $O(n^2)$, where n is the test length. However, it is much slower than Algorithm 20, because it is time-consuming to simulate input vector pairs and calculate *profit*.

10.4 Evaluation Results

10.4.1 Experimental Setup

The test generation framework, including the MERS core algorithms and the evaluation framework, is implemented using C. As shown in Fig. 10.1, the test generation framework can identify rare nodes, generate MERS testset, further tune the testset, and evaluate the effectiveness of testsets on random Trojans. We evaluated the approach on a subset of ISCAS-85 and ISCAS-89 benchmark circuits, as well as two large designs AES cipher and DLX processor [19]. The sequential circuits are converted into full scan mode. We also implemented the MERO [5] approach with parameter N of 1000 for comparison. The experiments were performed on a server with AMD Opteron Processor 6378 (2.4 GHz). The runtime for different benchmarks and different methods is shown in Table 10.1. The table also shows the number of rare nodes in each benchmark. We used 0.1 as the rare threshold to select rare nodes. We can see that if we use Algorithm 3 to reorder by Hamming distance, the runtime is about half of MERO on average. If we use Algorithm 4 to reorder by simulation, the runtime is about 7% longer on average. So it is reasonable to say that the generated testset is more effective than MERO given the similar test generation time.

10.4.2 Evaluation Criteria

When applying a testset to a circuit with Trojan, there are four criteria to evaluate the effectiveness of the testset:

- **AvgDeltaSwitch**: the average delta switch when applying the testset on this Trojan-infected circuit.
- **MaxDeltaSwitch**: the maximum delta switch when applying the testset.
- **AvgRelativeSwitch**: the average relative switch when applying the testset.
- **MaxRelativeSwitch**: the maximum relative switch when applying the testset. We choose this criterion as the **side-channel sensitivity** because this directly determines whether a Trojan can be detected through side-channel analysis.

AvgDeltaSwitch and *MaxDeltaSwitch* reflect the activity in Trojan, and *AvgRela-tiveSwitch* as *MaxRelativeSwitch* reflect the sensitivity of the side-channel signal in detecting the Trojan.

As for evaluation of testsets, we would expect a high-quality testset to have a good coverage over all possible Trojans. In the experiments, each testset is applied to 1000 randomly-inserted Trojan samples and these four values are computed for each testset. We would then take the average of these four metrics, which would reflect the capability of the testset to enable detection of different Trojans through side-channel analysis. The average $MaxRelativeSwitch$ would be most suitable for side-channel sensitivity evaluation, which is to maximize the sensitivity for an arbitrary Trojan in unknown circuit location.

10.4.3   Different Scan Modes

For sequential benchmarks, we assume that the sequential gates (i.e., the flip-flops) have full-scan capability during test. The initial states of the circuit can be set by the scan chain. Test vectors feed values to the primary inputs (PI) and the scan flip-flops (also called pseudo-PI). The controllability of the circuit states largely depends on the working mode of the scan chain. The transition test involves applying a vector pair (V_1, V_2) to the circuit. The first vector is to launch the circuit into a desired state. The transitions will be captured after V_2 is applied. V_1 will set the PI values as well as the initial states of circuit through SFFs. V_2 will feed the circuit with a different set of PI values. V_2 may or may not feed the SFFs with new values depending on the scan mode. We measure the number of switching in the circuit for side-channel analysis after V2 is applied.

A conventional scan chain can work in Launch-on-Shift (LoS) and Launch-on-Capture (LoC) modes [20]. In both of these two modes, V_2 only feeds the circuit with new values to PIs. The flip-flops will have values either directly from V_1 (shift by 1) or after propagating for one clock cycle. For LoS mode, the second vector V_2 is immediately applied after V_1 is shifted into SFFs. For LoC mode, the second vector V_2 waits for one clock cycle after V_1 is applied to the circuit. An enhanced scan chain can work in *Enhanced* mode [28, 31]. Compared to the LoS and LoC, the enhanced scan chain has one extra redundant flip-flop attached to each of the SFF. After the shifting process, the SFFs hold states for V_1 and the redundant FFs hold states for V_2. This feature enables both V_1 and V_2 to feed arbitrary values to the sequential gates. It comes at the cost of doubling the number of flip-flops. However, it provides high controllability and testability into the sequential circuits. Unless explicitly specified, the experiments assume that the enhanced scan mode is used for the sequential benchmarks.

10.4.4 Exploration of N

Figure 10.3 shows the distribution of MaxDeltaSwitch over 1000 random 8-trigger Trojan samples for two ISCAS-85 benchmarks. We choose different N to generate MERS testsets, to compare with the Random (10 K vectors) testset. For each testset, the box plot shows (minimum, first quartile, median, third quartile, maximum) values of MaxDeltaSwitch of the 1000 Trojan samples. It is clear from these plots that the distribution of MaxDeltaSwitch is constantly improving with increasing N. For $c2670$, the average MaxDeltaSwitch (as shown by the red lines) can reach 18.67 for MERS ($N = 1000$), while Random testset can achieve only 12.15. For $c3540$, the average MaxDeltaSwitch can reach 11.13 for MERS ($N = 1000$), while for Random testset it is only 9.19. The fact that the quality of MERS tests improves with increasing N is not surprising. It is similar to N-detect tests for stuck-at faults, where fault coverage is expected to improve with increasing N. The testset size also increases with N. The sizes of testsets for MERS ($N = 10, 20, 50, 100, 200, 500,$

Fig. 10.3 Impact of N (number of times that a rare node has rare switching) on MaxDeltaSwitch for benchmarks (**a**) c2670 and (**b**) c3540

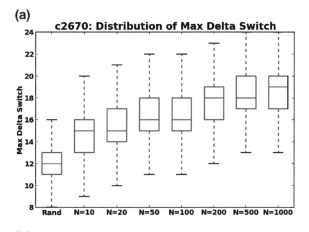

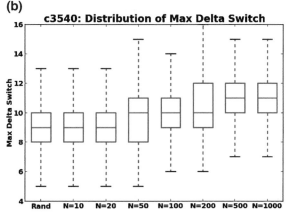

1000) are (71, 140, 347, 656, 1262, 3142, 6199) for $c2670$, and (161, 302, 742, 1441, 2858, 7070, 14250) for $c3540$. In most of the experiments, we choose a value of $N = 1000$, which is a good balance between testset quality and testset size. For fair comparison with Random testset, we will only take the first 10 K vectors of MERS testset if it is larger than 10 K.

10.4.5   Effect of Increased Total Switching

Figure 10.4 shows the average $MaxDeltaSwitch$ and the average $TotalSwitch$ of the testsets for 1000 8-trigger Trojan samples for different values of N. For both of the two benchmarks, the average $TotalSwitch$ increases with N as well as the average $MaxDeltaSwitch$. It is obvious that all the MERS testsets have much larger average $TotalSwitch$, compared with the Random testset. For $c2670$, the average $TotalSwitch$ for MERS ($N = 1000$) is 644.9, which is about 1.25X

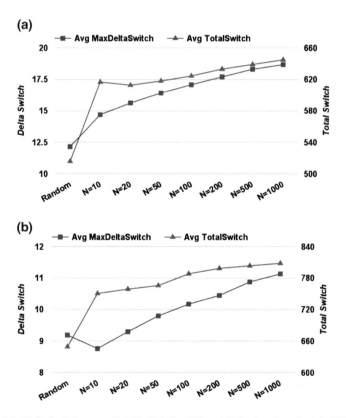

Fig. 10.4   MaxDeltaSwitch versus TotalSwitch for different N for benchmarks (**a**) c2670 and (**b**) c3540. MERS creates more switching in Trojan, as well as increased total switching

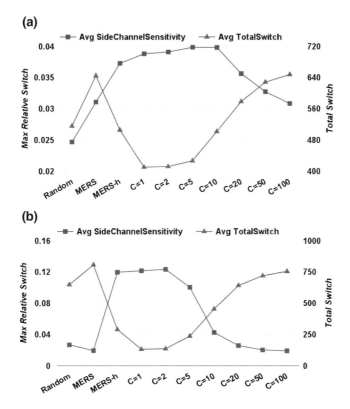

Fig. 10.5 Side-channel sensitivity versus $TotalSwitch$ for random, MERS, MERS-h, and MERS-s (with different C) for benchmarks (**a**) c2670 and (**b**) c3540

times of that of the Random testset (515.7). For $c3540$, the average $TotalSwitch$ for MERS ($N = 1000$) is 808, while Random testset is only 649.2. The insight that we can get from here is that MERS tends to increase the $TotalSwitch$ of the circuit, although it is designed to increase switches in rare nodes. The following subsection will show that the proposed reordering methods would be effective to reduce $TotalSwitch$ and thus increase side-channel sensitivity.

10.4.6   Effect of Weight Ratio (C)

The effectiveness of the two reordering methods can be observed in Figs. 10.5 and 10.6. As shown in Fig. 10.5, MERS-h can reduce $TotalSwitch$ and thus increase the relative switching (i.e., the side-channel sensitivity), compared with the original MERS testset. For MERS-s with different weight ratio C, side-channel sensitivity improves steadily with a small C, and then goes down when C is too

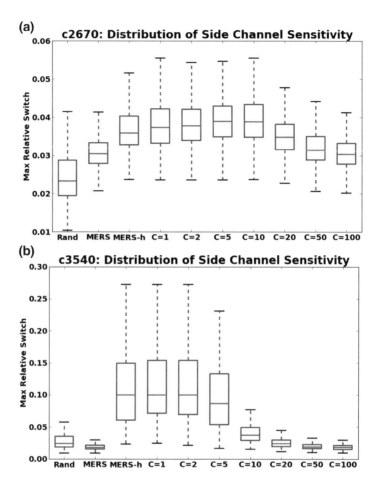

Fig. 10.6 Distribution of side-channel sensitivity for random, the original MERS, MERS-h, and MERS-s (with different C) for benchmarks (**a**) c2670 and (**b**) c3540

large. As the weight ratio tries to balance $DeltaSwitch$ and $TotalSwitch$, a large C will outweigh the influence of $TotalSwitch$, which will make it less different from the original MERS testset. In the following experiments, we choose the weight ratio as $C = 5$, as it provides a good balance between the total switching and rare switching.

Figure 10.6 shows detailed distribution of side-channel sensitivity for 1000 8-trigger Trojan samples with different choices of C. The reordering methods are working well to improve side-channel sensitivity, which is built on the fact that the original MERS testset is already of high quality in terms of $DeltaSwitch$, or switching in Trojans.

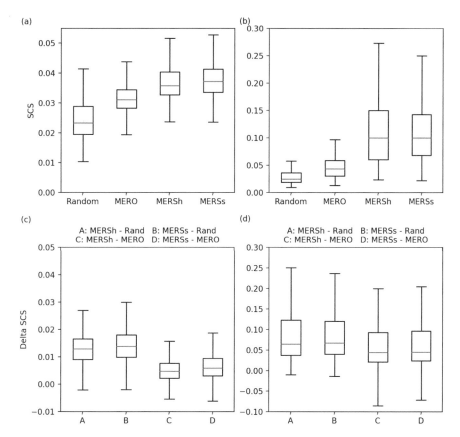

Fig. 10.7 Distribution of side-channel sensitivity (SCS) and delta SCS (1000 Trojan samples) compared with random testsets and MERO [5]. (**a**) C260: distribution of SCS. (**b**) C3540: distribution of SCS. (**c**) C260: distribution of delta SCS. (**d**) C3540: distribution of delta SCS

10.4.7 Increase in Trojan Activity

Figure 10.7 shows the distribution of the change in side-channel sensitivity for two benchmarks, compared with Random testset and MERO [5]. In Fig. 10.7a, b, we can see that the two approaches (MERS-h and MERS-s) can greatly improve the SCS compared with Random testsets as well as MERO [5]. Figure 10.7c, d show the Delta SCS when we look at each Trojan. Table 10.2 summarizes the percentage of Trojans whose SCS increased with the proposed approaches. Compared with Random testsets, more than 96.2% Trojans (for 1000 samples) have higher SCS with the proposed approaches. Compared with MERO testsets, more than 89.1% Trojans have higher SCS with the proposed approaches.

Table 10.3 shows that MERS (N=1000) is very effective in creating *DeltaSwitch* caused by arbitrary Trojans due to its statistical nature. The average

Table 10.2   Percent of Trojans (for 1000 Trojan samples) with SCS increased

Benchmark	MERSh-Rand	MERSs-Rand	MERSh-MERO	MERSs-MERO
c2670	96.2%	97.1%	89.1%	91.2%
c3540	99.8%	99.5%	92.3%	93.4%

Table 10.3   Comparison of MERS ($N = 1000$) with Random (10 K) for average MaxDeltaSwitch and average AvgDeltaSwitch, over 1000 random 8-trigger Trojans

Benchmark	Average MaxDeltaSwitch			Average AvgDeltaSwitch		
	Random	MERS	Improv.	Random	MERS	Improv.
c2670	12.15	18.67	53.67%	1.4289	6.8561	379.83%
c3540	9.19	11.13	21.16%	1.3716	2.9058	111.85%
c5315	9.51	13.80	45.16%	1.3116	3.9300	199.64%
c6288	6.63	7.26	9.63%	1.0636	4.8448	355.50%
c7552	8.53	12.00	40.76%	1.3488	2.7700	105.36%
s13207	6.63	8.83	33.18%	0.6428	0.9771	52.01%
s15850	7.53	10.84	43.99%	0.7465	1.3609	82.29%
s35932	15.16	15.37	1.35%	2.1803	6.8060	212.16%
Avg. improv.	–	–	31.11%	–	–	187.33%

Table 10.4   Comparison of MERS ($N = 1000$) with Random (10 K) for average *MaxRelativeSwitch* (side-channel sensitivity) and average *AvgRelativeSwitch*, over 1000 random samples of 8-trigger Trojans

Benchmark	Average MaxRelativeSwitch (Side-Channel Sensitivity)			Average AvgRelativeSwitch		
	Random	MERS	Improv.	Random	MERS	Improv.
c2670	0.02469	0.03108	25.90%	0.00255	0.01054	314.14%
c3540	0.02670	0.01933	−27.59%	0.00214	0.00361	69.12%
c5315	0.00526	0.00766	45.72%	0.00075	0.00200	165.65%
c6288	0.00534	0.00395	−26.06%	0.00059	0.00219	270.68%
c7552	0.00452	0.00852	88.48%	0.00058	0.00113	94.65%
s13207	0.00756	0.00844	11.64%	0.00066	0.00085	28.22%
s15850	0.00593	0.00716	20.70%	0.00053	0.00082	54.25%
s35932	0.00523	0.00587	12.29%	0.00060	0.00223	268.54%
Avg. improv.	–	–	18.89%	–	–	158.16%

Max Delta Switch increases by 31.11% and the average *Avg Delta Switch* by 187.33% on average for different benchmarks, compared with Random testset. This shows the effectiveness of MERS in creating Trojan activity.

Table 10.4 shows that MERS is also helpful in improving RelativeSwitch. The average AvgRelativeSwitch increased by 158.16%, compared with Random testsets. For average MaxRelativeSwitch (side-channel sensitivity), MERS has an average improvement of 18.89%. However, side-channel sensitivity values for benchmark $c3540$ and $c6288$ are not as good as those of Random testsets. This is due to the fact that MERS testset also increases the total switching, when it is making efforts to cause rare nodes switching. This phenomenon is illustrated and explained in Figs. 10.4 and 10.5, and this side effect can be improved by the two reordering algorithms as shown in Tables 10.5 and 10.6.

10.4.8   Side-Channel Sensitivity Improvement

To this point, we have explored the parameters: N for MERS and C for MERS-s. We choose $N = 1000$ and $C = 5$ in the following experiment to compare the proposed schemes with Random testset and MERO. Tables 10.5 and 10.6 show the improvement of proposed approaches on side-channel sensitivity for 4-trigger and 8-trigger Trojans.

Table 10.5 shows that MERS, MERS-h, and MERS-s have 10.37%, 138.44%, and 152.26% improvement over the Random testsets, respectively. While the original MERS testsets is 23.95% worse than MERO testsets, MERS-h and MERS-s have 52.62% and 62.01% improvement over MERO. Table 10.6 shows the results for 8-trigger Trojans. Compared to Random testsets, MERS, MERS-h, and MERS-s can have 18.89%, 107.53%, and 96.61% improvement, respectively. The original MERS testsets is 12.43% worse than MERO testsets. MERS-h and MERS-s testsets can improve the side-channel sensitivity by 40.79% and 38.50%, respectively.

In this section, we explore the impact of different values of N for MERS and observe the effectiveness of MERS to maximize Trojan activity as N increases. We confirm the superiority of MERS testsets over Random testsets in Sect. 10.4.7 on creating switching activity in randomly sampled Trojans. We observed that the total switching was also likely to increase while MERS made efforts to maximize rare switching in Trojans. The two reordering methods (MERS-h and MERS-s) successfully had the total switching under control while maintaining the rare switching high.

10.5   Scalability to Large Designs

In this section, we investigate the scalability of the proposed approach to large designs. We compare the controllability of different scan modes and their effects on side-channel sensitivity. We apply region-based MERS on the three sequential ISCAS benchmarks and side-channel sensitivity can improve as we divide the design into more regions. We also generate tests for two large benchmarks (AES cipher and DLX processor) from OpenCores and design partitioning can significantly improve the side-channel sensitivity.

Table 10.5 Comparison of average **side-channel sensitivity** between random (10 K), MERO, and MERS testsets, $N = 1000$, $C = 5$ for MERS-s, over 1000 random samples of 4-trigger Trojans

Benchmark	Comparison testsets		Proposed schemes			Improvement to random			Improvement to MERO		
	Random	MERO	MERS	MERS-h	MERS-s	MERS	MERS-h	MERS-s	MERS	MERS-h	MERS-s
c2670	0.01703	0.02571	0.02231	0.03035	0.03308	31.01%	78.27%	94.31%	−13.23%	18.07%	28.69%
c3540	0.02144	0.04238	0.01336	0.10677	0.11067	−37.71%	397.97%	416.16%	−68.48%	151.96%	161.16%
c5315	0.00445	0.01082	0.00747	0.01287	0.01586	67.79%	188.97%	256.29%	−30.97%	18.89%	46.59%
c6288	0.00480	0.00395	0.00313	0.00741	0.00896	−34.81%	54.47%	86.85%	−20.88%	87.50%	126.80%
c7552	0.00351	0.00737	0.00491	0.01250	0.01168	39.61%	255.63%	232.38%	−33.46%	69.50%	58.42%
s13207	0.00568	0.00617	0.00619	0.00773	0.00826	9.07%	36.24%	45.49%	0.31%	25.29%	33.80%
s15850	0.00447	0.00487	0.00474	0.00691	0.00634	6.14%	54.83%	42.06%	−2.75%	41.86%	30.17%
s35932	0.00354	0.00463	0.00361	0.00500	0.00512	1.89%	41.17%	44.53%	−22.12%	7.90%	10.48%
Avg. improve.	–	–	–	–	–	10.37%	138.44%	152.26%	−23.95%	52.62%	62.01%

Table 10.6 Comparison of average **side-channel sensitivity** between random (10 K), MERO, and MERS testsets, $N = 1000$, $C = 5$ for MERS-s, over 1000 random samples of 8-trigger Trojans

Benchmark	Comparison testsets		Proposed schemes			Improvement to random			Improvement to MERO		
	Random	MERO	MERS	MERS-h	MERS-s	MERS	MERS-h	MERS-s	MERS	MERS-h	MERS-s
c2670	0.02469	0.03204	0.03108	0.03729	0.03984	25.90%	51.05%	61.40%	−3.01%	16.37%	24.35%
c3540	0.02670	0.05532	0.01933	0.11974	0.10037	−27.59%	348.53%	275.96%	−65.05%	116.47%	81.44%
c5315	0.00526	0.00875	0.00766	0.01020	0.01129	45.72%	94.03%	114.78%	−12.38%	16.66%	29.14%
c6288	0.00534	0.00412	0.00395	0.00649	0.00790	−26.06%	21.55%	47.97%	−4.20%	57.49%	91.72%
c7552	0.00452	0.00914	0.00852	0.01437	0.01149	88.48%	217.78%	154.00%	−6.70%	57.31%	25.74%
s13207	0.00756	0.00838	0.00844	0.01053	0.01112	11.64%	39.24%	47.05%	0.69%	25.58%	32.63%
s15850	0.00593	0.00722	0.00716	0.00923	0.00818	20.70%	55.69%	37.94%	−0.87%	27.86%	13.28%
s35932	0.00523	0.00638	0.00587	0.00692	0.00700	12.29%	32.39%	33.80%	−7.90%	8.58%	9.74%
Avg. improve.	–	–	–	–	–	18.89%	107.53%	96.61%	−12.43%	40.79%	38.50%

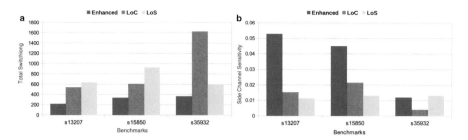

Fig. 10.8 Comparison of Enhanced, LoC, and LoS modes. (**a**) Total switching, (**b**) side-channel sensitivity

10.5.1   Controllability of Different Scan Modes

The scan modes have direct influence on the effectiveness of the region-based MERS approach. Figure 10.8 shows the *Total Switching* and the *side-channel sensitivity* when different scan modes are used for region-based MERS approach (each benchmark has four regions in this example). In Fig. 10.8a, the Enhanced mode can greatly reduce the *Total Switching* compared with LoC and LoS. In Fig. 10.8b, the Enhanced mode can greatly improve the *side-channel sensitivity* compared with LoC and LoS.

There are two factors that enable the *Enhanced* mode to do much better than the LoC and LoS modes. (1) We try to reduce the background switching by assigning 0's to the flip-flops that are outside of the targeted region. For the *Enhanced* mode, we assign 0's to both V_1 and V_2 for those flip-flops outside of the targeted region. This enables the *Enhanced* mode to have full controllability to turn the other regions "dark." For the LoC and LoS modes, we assign 0's to V_1 for those flip-flops that are outside of the targeted region, while V_2 cannot directly assign values to flip-flops. For LoC mode, the flip-flop states before capture will be the states after the circuit propagates one cycle after V_1. For LoS mode, the flip-flop states will be shifted by 1 from V_1. (2) The *Enhanced* mode has the benefit of using V_2 to assign arbitrary values to the flip-flops inside the targeted region. In contrast, LoC has no direct control over the states after one clock cycle and LoS has to assign the in-region flip-flops to V1 shifted by 1. In MERS (Algorithm 19), we mutate both the PIs and the pseudo-PIs (i.e., the values for the in-region flip-flops) to generate high-quality test for each region. Under the *Enhanced* mode, we can mutate the vector V_2 to find beneficiary values for pseudo-PIs.

10.5.2   Effectiveness of Design Partitioning

Figure 10.9 shows the results for region-based MERS approach on the three sequential benchmarks. We compare the MERS testsets produced by 1 region, 2 regions, 4 regions, and 8 regions. The *Total Switching* and *side-channel sensitivity*

Fig. 10.9 (**a**) *Total switching.*
(**b**) *Side-channel sensitivity*
for in-region Trojans samples.
(**c**) *Side-channel sensitivity*
for cross-region Trojans

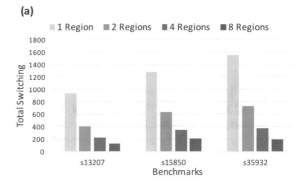

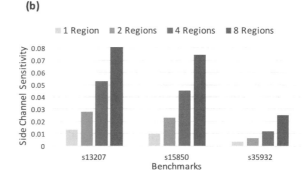

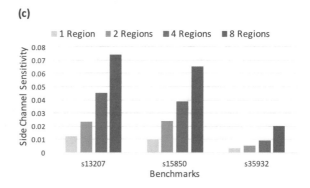

numbers are the averaged values of 1000 random in-region (out-region) Trojans. Here an in-region Trojan means that its trigger edges and payload edge belong to the same region. We have 125 random in-region Trojans from each of the 8 regions to form a set of 1000 random Trojans for each benchmark.

As shown in Fig. 10.9a, the averaged *Total Switching* decreases drastically as we partition the design into more regions. As the number of regions doubles, the averaged *Total Switching* reduces almost by half. MERS with 8 regions can reduce the *Total Switching* by 7.40X for s13207, 6.28X for s15850, and 8.12X for s35932, compared to MERS with only 1 region. As shown in Fig. 10.9b, *side-channel sensitivity* improves significantly as the number of regions increases for in-region Trojans. MERS with 8 regions can improve the *side-channel sensitivity* by 6.24X for s13207, 7.51X for s15850, and 7.49X for s35932, compared to MERS with only 1 region.

Figure 10.9c shows the average SCS of 1000 cross-region Trojan samples. We can still see the trend that SCS will greatly increase as we divide the design into more regions. We observe slightly lower SCS compared with in-region Trojan samples. For the benchmark s13207, cross-region Trojan samples have 16.1% less SCS for *2 Regions*, 14.0% less SCS for *4 Regions*, and 7.9% less SCS for *8 Regions*, compared with in-region Trojan samples. For the benchmark s35932, cross-region Trojan samples have 18.8% less SCS for *2 Regions*, 22.5% less SCS for *4 Regions*, and 20.3% less SCS for *8 Regions*, compared with in-region Trojan samples. Thus the conclusion is that cross-region Trojans can still significantly benefit from the proposed approach.

10.5.3 Test Generation for Large OpenCores Benchmarks

In this subsection, we apply the region-based MERS approach on two large designs (AES cipher and DLX processor). AES cipher has 15086 nodes and DLX processor has 18123 nodes. They are about three times as large as the largest ISCAS benchmark s35932. The results show that this approach is scalable for large designs. Direct application of MERS on AES takes about 7 days to generate and reorder tests, and about 9 days for DLX. After functionally partitioning AES into three regions, the largest region can finish in 4 days (we generate the tests for each region in parallel). After functional partitioning of DLX and structural partitioning of its decode module, we can finish the test generation and reordering for DLX in 3 days. In this part, we assume that the designs are equipped with enhanced scan chain, which provides us the most controllability for test generation.

10.5.3.1 AES

Figure 10.10a shows the abstracted representation of an AES cipher. We use functional partition to segment it into three regions. It has two obvious submodules: *Key Expansion* and *Round Permutation*, which we choose as two regions. The third region contains the rest of the circuit, which is mostly the control logic and input/output buffers.

Table 10.7 compares the side-channel sensitivity on 15 random Trojans for three testsets: Random, MERS (whole design), and MERS-FP (with functional partition). For the MERS and MERS-FP, we use the test generation detailed in Algorithm 19 and the test reordering detailed in Algorithm 20. Compared with the Random testset, the MERS testset can improve the side-channel sensitivity by only 5% on average.

Fig. 10.10 Design partition
for AES and DLX. (**a**)
Functional partition on AES.
(**b**) Functional partition on
DLX + structural partition on
decode

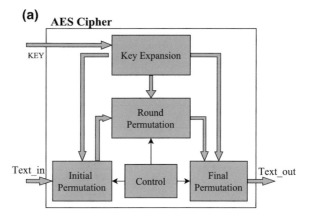

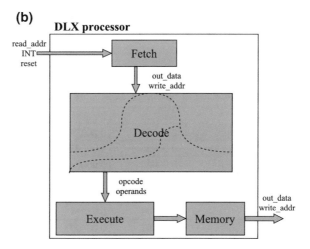

The MERS-FP testset can improve the side-channel sensitivity by 160% on average.
The functional partition significantly improved the side-channel sensitivity (about
2.6X) over the Random testset.

10.5.3.2   DLX

Figure 10.10b shows the abstracted representation of a DLX processor. We use
functional partition to segment it into four regions: *Fetch*, *Decode*, *Execute*, and
Memory. However, the Decode consumes majority of the chip area (accounting
for 71% of the whole design area). We used the hypergraph partitioning tool
hMETIS [15] to further partition the Decode region into four regions of roughly
equal size (in terms of number of gates/vertices).

Table 10.8 compares the side-channel sensitivity on 15 random Trojans for
four testsets: Random, MERS (whole design), MERS-FP (with functional parti-

Table 10.7 AES testsets: [Random, MERS, MERS-FP]

Testset	Random	MERS	MERS-FP
Trojan 1	0.00125	0.00154	0.00568
Trojan 2	0.00119	0.00135	0.00538
Trojan 3	0.00136	0.00143	0.00525
Trojan 4	0.00170	0.00151	0.00573
Trojan 5	0.00125	0.00167	0.00568
Trojan 6	0.00107	0.00132	0.00196
Trojan 7	0.00105	0.00143	0.00235
Trojan 8	0.00148	0.00158	0.00196
Trojan 9	0.00090	0.00083	0.00110
Trojan 10	0.00083	0.00107	0.00127
Trojan 11	0.00267	0.00237	0.00389
Trojan 12	0.00227	0.00255	0.00356
Trojan 13	0.00235	0.00226	0.00952
Trojan 14	0.00262	0.00235	0.00635
Trojan 15	0.00235	0.00228	0.00354
Average SCS	0.00162	0.00170	0.00421
Average improve.	–	5%	160%

Table 10.8 DLX testsets: [Random, MERS, MERS-FP, MERS-FP+SP]

Testset	Random	MERS	MERS-FP	MERS-FP+SP
Trojan 1	0.00045	0.00059	0.00059	0.00202
Trojan 2	0.00055	0.00067	0.00067	0.01515
Trojan 3	0.00066	0.00087	0.00087	0.00448
Trojan 4	0.00050	0.00062	0.00065	0.00448
Trojan 5	0.00079	0.00059	0.00059	0.00202
Trojan 6	0.00042	0.00085	0.00085	0.00285
Trojan 7	0.00061	0.00090	0.00090	0.00384
Trojan 8	0.00081	0.00099	0.00099	0.00632
Trojan 9	0.00051	0.00096	0.00096	0.00210
Trojan 10	0.00059	0.00086	0.00126	0.00673
Trojan 11	0.00045	0.00073	0.00078	0.02273
Trojan 12	0.00038	0.00067	0.00067	0.00202
Trojan 13	0.00075	0.00093	0.00093	0.00384
Trojan 14	0.00051	0.00079	0.00079	0.00210
Trojan 15	0.00050	0.00062	0.00062	0.01515
Average SCS	0.00056	0.00077	0.00081	0.00639
Average improve.	–	37%	43%	1033%

tion), MERS-FP+SP (with functional partition followed by structural partition). Compared with the Random testset, the MERS testset can improve the side-channel sensitivity by about 37% on average, and MERS-FP testset can improve by 43%. The MERS and MERS-FP have very close side-channel sensitivity numbers, because the Decode region is very huge. The MERS-FP+SP testset can significantly improve the side-channel sensitivity (about 11X times) over the Random testset. The experiments on AES and DLX have shown that this approach is scalable to large designs to greatly improve the side-channel sensitivity for hardware Trojan detection.

10.6 Summary

We have presented a framework for scalable test generation, called MERS, which can significantly improve the Trojan detection sensitivity in side-channel analysis based Trojan detection. The approach aims at statistically increasing switching activity in an unknown Trojan to amplify the Trojan effect in presence of large process variations. Such a test generation approach will, in general, be effective for any side-channel analysis approaches that rely on activity in Trojan circuits (e.g., transient current, dynamic power profile, or electromagnetic emanation based methods). MERS is effective for any Trojan forms/sizes, as long as a Trojan is implanted through alterations in a circuit structure—the most dominant mode of Trojan implantation. The simulation results on a set of benchmark circuits show that the proposed approach can significantly improve the side-channel sensitivity by 97%, compared with random tests for a large set of arbitrary Trojans. Furthermore, this approach is scalable to large designs (e.g., AES cipher and DLX processor), which can improve side-channel sensitivity by 1.6X times for AES, and 10X times for DLX. Further, the approach can work for different DFT configurations. The results demonstrated that a scalable statistical test generation can serve as an essential component in any side-channel analysis based hardware Trojan detection framework.

References

1. M. Banga, M. Hsiao, A region based approach for the identification of hardware Trojans, in *IEEE International Workshop on Hardware-Oriented Security and Trust (HOST)* (2008)
2. M. Banga, M. Chandrasekar, L. Fang, M. Hsiao, Guided test generation for isolation and detection of embedded Trojans in ICs, in *ACM Great Lakes Symposium on VLSI (GLSVLSI)* (2008), pp. 363–366
3. S. Borkar, T. Karnik, S. Narendra, J. Tschanz, A. Keshavarzi, V. De, Parameter variations and impact on circuits and microarchitecture, in *ACM/IEEE Design Automation Conference (DAC)* (2003), pp. 338–342
4. R. Chakraborty, S. Bhunia, Security against hardware Trojan through a novel application of design obfuscation, in *ACM International Conference on Computer-Aided Design (ICCAD)* (2009), pp. 113–116
5. R. Chakraborty, F. Wolff, S. Paul, C. Papachristou, S. Bhunia, MERO: a statistical approach for hardware Trojan detection, in *International Workshop on Cryptographic Hardware and Embedded Systems* (2009), pp. 396–410

6. J. Cruz, Y. Huang, P. Mishra, S. Bhunia, An automated configurable Trojan insertion framework for dynamic trust benchmarks, in *Design Automation and Test in Europe (DATE)*, Dresden, Germany, March 19–23 (2018)
7. J. Cruz, P. Mishra, S.Bhunia, The metric matters: how to measure trust, in *Design Automation Conference (DAC)*, Las Vegas, June 2–6 (2019)
8. D. Du, S. Narasimhan, R. Chakraborty, S. Bhunia, Self-referencing: a scalable side-channel approach for hardware Trojan detection, in *International Workshop on Cryptographic Hardware and Embedded Systems (CHES)* (2010), pp. 173–187
9. S. Dupuis, P. Ba, G. Natale, M. Flottes, B. Rouzeyre, A novel hardware logic encryption technique for thwarting illegal overproduction and hardware Trojans, in *IEEE 20th International On-Line Testing Symposium (IOLTS)* (2014), pp. 49–54
10. F. Farahmandi, Y. Huang, P. Mishra, Trojan localization using symbolic algebra, in *Asia and South Pacific Design Automation Conference (ASPDAC)* (2017), pp. 591–597
11. Y. Huang, P. Mishra, Trace buffer attack on the AES cipher. J. Hardw. Syst. Secur. **1**(1), 68–84 (2017)
12. Y. Huang, S. Bhunia, P. Mishra, MERS: statistical test generation for side-channel analysis based Trojan detection, in *ACM Conference on Computer and Communications Security (CCS)* (2016), pp. 130–141
13. Y. Huang, S. Bhunia, P. Mishra, Scalable test generation for Trojan detection using side channel analysis. IEEE Trans. Inf. Forensics Secur. **13**(11), 2746–2760 (2018)
14. Y. Jin, Y. Makris, Hardware Trojan detection using path delay fingerprint, in *IEEE International Workshop on Hardware-Oriented Security and Trust (HOST)* (2008)
15. G. Karypis, R. Aggarwal, V. Kumar, S. Shekhar, Multilevel hypergraph partitioning: applications in VLSI domain. IEEE Trans. Very Large Scale Integr. Syst. **7**(1), 69–79 (1999)
16. Y. Lyu, P. Mishra, A survey of side channel attacks on caches and countermeasures. J. Hardw. Syst. Secur. **2**, 33–50 (2018)
17. Y. Lyu, P. Mishra, Efficient test generation for Trojan detection using side channel analysis, in *Design Automation and Test in Europe (DATE)*, Florence, Italy, March 25–29 (2019)
18. P. Mishra, S. Bhunia, M. Tehranipoor (eds.) *Hardware IP Security and Trust*. Springer, Basel (2017). ISBN 9783319490250
19. OpenCores, Project *aes_core* and *dlx*. http://www.opencores.org
20. I. Park, E.J. McCluskey, Launch-on-shift-capture transition tests, in *IEEE International Test Conference*, Santa Clara, CA (2008), pp. 1–9
21. I. Pomeranz, S. Reddy, A measure of quality for n-detection test sets. IEEE Trans. Comput. **53**(11), 1497–1503 (2004)
22. R. Rad, J. Plusquellic, M. Tehranipoor, A sensitivity analysis of power signal methods for detecting hardware Trojans under real process and environmental conditions. IEEE Trans. Very Large Scale Integr. Syst. **18**(12), 1735–1744 (2010)
23. J. Rajendran, Y. Pino, O. Sinanoglu, R. Karri, Security analysis of logic obfuscation, in *ACM/IEEE Design Automation Conference (DAC)* (2012), pp. 83–89
24. S. Saha, R. Chakraborty, S. Nuthakki, Anshul, D. Mukhopadhyay, Improved test pattern generation for hardware Trojan detection using genetic algorithm and boolean satisfiability, in *International Workshop on Cryptographic Hardware and Embedded Systems* (2015), pp. 577–596
25. H. Salmani, M. Tehranipoor, Layout-aware switching activity localization to enhance hardware Trojan detection. IEEE Trans. Inf. Forensics Secur. **7**(1), 76–87 (2012)
26. H. Salmani, M. Tehranipoor, J. Plusquellic, A novel technique for improving hardware Trojan detection and reducing Trojan activation time. IEEE Trans. Very Large Scale Integr. Syst. **20**(1), 112–125 (2012)
27. P. Subramanyan, S. Ray, S. Malik, Evaluating the security of logic encryption algorithms, in *IEEE International Symposium on Hardware Oriented Security and Trust (HOST)* (2015), pp. 137–143

28. A.K. Suhag, V. Shrivastava, Delay testable enhanced scan flip-flop: DFT for high fault coverage, in *International Symposium on Electronic System Design*, Kochi, Kerala (2011), pp. 129–133

29. S. Wei, M. Potkonjak, Scalable hardware Trojan diagnosis. IEEE Trans. Very Large Scale Integr. Syst. **20**(6), 1049–1057 (2012)

30. F. Wolff, C. Papachristou, S. Bhunia, R.S. Chakraborty, Towards Trojan-free trusted ICs: problem analysis and detection scheme, in *Design, Automation and Test in Europe (DATE)* (2008), pp. 1362–1365

31. G. Xu, A.D. Singh, Low cost launch-on-shift delay test with slow scan enable, in *IEEE European Test Symposium (ETS'06)*, Southampton (2006), pp. 9–14

32. B. Zhou, W. Zhang, S. Thambipillai, J. Teo, A low cost acceleration method for hardware Trojan detection based on fan-out cone analysis, in *ACM International Conference on Hardware Software Codesign and System Synthesis* (2014), p. 28

Chapter 11
Hardware Trojan Detection Schemes Using Path Delay and Side-Channel Analysis

11.1 Introduction

Hardware Trojans (HT) are deliberate and malicious changes to an electronic device that adds or removes functionality or reduces reliability of an integrated circuit (IC), printed circuit board (PCB), or system [6–9, 14, 23, 27, 52, 54]. The changes can be designed to leak secret information, e.g., encryption keys or other types of private internal information, or they may be designed to cause the system to fail at some specific or predetermined time while the IC is in mission mode. The business model of distributed and outsourced design, integration, manufacturing, packaging, and distribution channels open up challenges such as intellectual property (IP) piracy, reverse engineering of netlist from GDSII, integrated circuit (IC) cloning, counterfeit attacks, and Trojan insertions. The shrinking integrated circuit feature size and increased gate density per wafer have been made possible with the advancements in photolithography techniques; however, this has caused the manufacturing processes to become very complex and the cost reaching billions of dollars. With the rapidly improving technology, the fabrication plant requires high operating and maintenance cost; therefore, the business model of outsourcing and off-shoring production process is observed in the leading semiconductor industry for cost reduction. Additionally, because of increased complexity and integration of system design, nearly every step of the modern design process, from architecture, through RTL, layout, split manufacturing [46], packaging, IC testing, distribution, and system integration is 'farmed out' to individual companies located all over the world. This distributed IC production business and off-shoring of the manufacturing operations and IC designers having less or no longer control on fabrication process are becoming an important driver of emerging security and trust problems [53]. Adversaries design the HT to be difficult to discover, either accidentally via manufacturing test or purposely using tests specifically designed to activate the HT.

Furthermore, in recent intellectual property IP-reuse based design flow in system on chip requires additional IP protection schemes to avoid illegal modifications,

© Springer Nature Switzerland AG 2020
F. Farahmandi et al., *System-on-Chip Security*,
https://doi.org/10.1007/978-3-030-30596-3_11

piracy, and ownership issues. IPs can come in form of soft, firm, or hard IPs. For example, IP reuse in the soft form are the synthesizable register-transfer-level (RTL) description, firm IPs are the gate-level designs to integrate with the firmware and hard IP is distributed in the form of GDSII design database. In this design approach, IP are transparent at system design level, manufacturing facility, and distribution chain, making it susceptible to security and privacy attacks at different entry levels. Sophisticated HT insertion strategies consider resilience to advanced HT detection methods that utilize high-resolution measurements of side-channel signals, such as electromagnetic (EM) emanations, power consumption (steady-state IDDQ or transient IDDT), delay testing, and temperature profiling. In addition to these testing challenges, HT detection methods are further tasked to deal with several other fundamental HT properties. First, the task of identifying an HT is akin to finding a needle in a haystack, i.e., the adversary has a huge advantage because he/she can choose to insert the HT anywhere while the trusted authority is tasked with determining if the IC has in fact been modified and if so, finding the unknown malicious function in a "sea" of gates. Second, HT and "bugs," either hardware or software, share the same characteristics, and it is widely accepted that finding all the bugs in a complex program is generally infeasible. In fact, cleverly inserted HT can be designed to appear as bugs, making it difficult to decide if the malicious function, if discovered, was accidental or purposeful. Third, any attempt by the trusted authority to increase the "ease" of HT detection may be visible to the adversary, i.e., the adversary can reverse engineer the IC and avoid countermeasures added by the trusted authority. Fourth, the adversary can choose to "selectively" insert the HT into only a subset of the manufactured ICs, making it necessary to verify all manufactured ICs. Last, HT designed to leak information may not cause a change in the functional behavior of the IC, and, therefore, the trusted authority may need to apply non-standard tests, e.g., tests for anomalous EM radiations. Moreover, the appropriate detection strategy will vary greatly depending on the assumptions made regarding the "insertion point," i.e., design-inserted HT requires very different detection techniques than those inserted into a layout description of the design.

The only advantage afforded to the trusted authority is that his/her detection strategy can be "parallelized" because the HT needs to be detected only once and is, in most cases because of mask cost issues, inserted in the same fashion in every copy of the targeted IC population. Therefore, tests applied post-manufacturing can be partitioned among multiple independent IC testers (referred to as automatic test equipment or ATE) and applied in parallel. Unfortunately, even high levels of parallelism "run out of gas" when the full extent of the search space, both combinational and sequential, is considered.

11.1.1 Threat Models

IC piracy and tempering the design for malicious objectives is a major security concern in the trending business model of IP reuse and offshore manufacturing. It is very important to understand different attack models and piracy act:

1. Reverse engineering: GDSII can be used to reverse engineer the netlist and interpret the functionality to steal it.
2. Clones: An attacker in the system design flow can steal the IP or IC and with a few modifications, claim the ownership and make illegal copies.

   Overbuilding: Mass production of ICs from same masks reduces considerable cost of fabricating extra chips and selling them in black. Without integration of special techniques, identification of the individual parts is a challenge. Thus good chips cannot be separated from the overbuild chips with the current design flow.
3. Counterfeit chips: Counterfeit chips are intended to deceptively represent an authentic component that could be recycled or cloned chip.
4. Repudiation: In case of counterfeit chip detection, adversary can refuse the reasonability.
5. Side-channel attacks: Key based security in designs can be broken by the side-channel attacks. Advance active hardware obfuscation techniques require keys to unlock functionality. Once the key is compromised, adversary can reverse engineer the original netlist and remove all the keys, defeating the whole security through obscurity.
6. Trojan detection: On successfully reverse engineering the design, adversary can include Trojans, hidden malicious modifications to the circuitry. A Trojan payload can be activated during the life cycle of IC without the knowledge of user.

This chapter is specifically focused on surveying methods that utilize very precise analog based testing to discover HT. The underlying basis of these methods can be characterized by the Heisenberg principle or observer effect, i.e., any attempt to measure or monitor a system changes its behavior. The testing methods described herein attempt to determine if an adversary has inserted an HT that is "observing" the evolving state of the IC, which is used by the adversary as the mechanism to activate the HT. In particular, we survey path-delay-based testing methods which are designed to detect subtle changes in delay introduced by the HT connections and gate insertions, referred to as the trigger and payload of the HT, respectively.

The authors of [59] propose a generic characterization of these concepts as shown in Fig. 11.1. The rest of this chapter is organized as follows. Section 11.2 presents a high-level view of HT insertion strategies and discusses the constraints on the detection methods. Section 11.3 covers HT detection strategies designed to detect layout or GDSII Trojans (other HT insertion points are detailed in other chapters of this book) with subsections that survey detection methods that analyze "side-channel" signals, e.g., power and delay. Section 11.4 describes important fundamental concepts related to implementing path-delay-based HT methods. Sec-

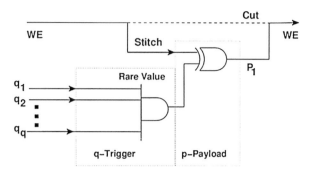

Fig. 11.1 Generic characterization of a hardware Trojan trigger and payload from [46]. Trigger signals $q1$ through qq typically connect to nodes in the existing design and therefore add capacitive load to these signals, creating an observer effect. Both the trigger signals and payload add delay to paths in the existing design

tion 11.5 provides a survey of delay-based HT detection techniques, while Sect. 11.6 describes the proposed multiple-parameter side-channel techniques. Conclusions are provided in Sect. 11.7.

11.2   Hardware Trojan Insertion

The horizontal dissemination of the IC design, fabrication, and test processes to many distinct companies around the world has dramatically increased the potential for malicious activities. Intellectual property (IP) block reuse has compounded this threat by partitioning the design space itself among multiple third-party vendors. Standardization activities have enabled multiple independently designed IP blocks to seamlessly integrate into CAD tool flows. However, the electronic design automation (EDA) community developed this multi-party collaborative design system using a model in which all parties are largely trusted. Unfortunately, the same types of malicious activities endured by the software community are now presenting themselves in the hardware design community.

All of the primary processes associated with design, manufacture, and test are vulnerable to malicious activities where adversaries can add to, remove from, or change the functionality of the IC. We refer to these opportunities as insertion points. Figure 11.2 provides a graphical illustration of the major insertion points, which are further distinguished by the following list:

- Designing third-party IP blocks
- Developing CAD tool scripts
- Integration activities where IP blocks and glue logic are assembled into system-on-chip (SoC) ICs
- Behavior synthesis and place and route (PnR) carried out by CAD tools

Fig. 11.2 Hardware Trojan insertion points

- Layout mask data generation and mask preparation
- Process parameter control mechanism used in the multistep fabrication process
- Supply chain transactions associated with transferring wafers from one facility to another
- Generating test vectors using automatic test pattern generation (ATPG)
- Wafer-probing activities associated with measuring test structures and detecting defects
- Supply chain transactions associated with creating and transferring dice
- Processes responsible for packaging ICs
- Applying ATPG vectors to packaged ICs using ATE
- Supply chain transactions associated with transferring packaged parts
- Printed circuit board (PCB) design and fabrication
- Processes responsible for installing PCB components (populating PCBs)
- Supply chain transactions associated with transferring boards
- System integration and deployment activities

The wide range and widely distributed nature of these activities presents an overwhelming opportunity for subversion. Moreover, the wide diversity among the tasks will require a very sophisticated and complex system to manage the entire set of trust vulnerabilities from start to finish. The research community is tackling the trust challenges one at a time and is focused on those that are the most attractive insertion points for adversaries. For example, subversion of IP blocks is a serious concern given the ease in which malicious functionalities can be covertly inserted and the absence of alternate representations and models to which the IPs can be compared [37]. Layout modifications and IC fabrication insertion points represent another important focus area, especially given the huge complexity associated with analyzing fabricated ICs at this lowest layer of design abstraction, and the wide range of opportunities available to the adversary in designing HT with sophisticated, sometime analog, triggering and payload mechanisms. Note that significant differences exist in the HT countermeasures and detection strategies that are applicable even when only considering these two insertion points.

For example, golden models are not available at the IP block insertion point, but architectural changes that obfuscate the design are available as countermeasures. On the other hand, the layout insertion point allows layout design data to be used to validate the functional and analog behaviors of the IC, but obfuscation is limited to "dummy via" insertion and other nano-level manipulations of the design. Also, side-channel information is not available or is not accurate enough to be useful for IP blocks but can be leveraged as a very powerful HT detection method for layout-level validation. The focus of this chapter is on HT detection methods, and countermeasures where appropriate, that are applicable at the layout level. Other chapters of this text survey techniques which target other insertion points.

11.3   Approaches to Detect Layout-Inserted Hardware Trojans

A layout is a physical representation of the design, i.e., it is a set of geometric shapes that rep-resent a physical model of the IC. The shapes define transistors, wires, vias, and contacts. A layout is the lowest layer of abstraction in the design process and contains all the logic gates that define the function as well as all the electrical connections between the logic gates and the power supply rails. The complexity of layouts increases as technology feature sizes shrink into the nanometer regime, and additional wiring layers are added. Figure 11.3 shows several standard (std.) cell layouts on the left and a tool-synthesized layout of a relatively small functional unit called the Advanced Encryption Standard (AES). The layout of the AES IP block contains approx. 12,000 std. cells and 50,000 wires and typically would represent

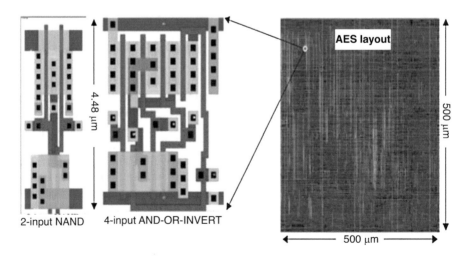

Fig. 11.3 Layout of std. cells (left) and AES layout (right)

one IP block of several 100 on a modern SoC. The technology used in this example is an IBM 90 nm process which provides nine vertically stacked layers for metal wires. The image is a screen capture of the designer's view of the layout using Cadence Virtuoso. Most layout design tools provide this type of top-down view, with upper metal layers obscuring the transistors in the bottom-most layer, i.e., nearly all of what is shown in the AES layout are metal wires.

Once the physical model of the layout is completed as shown by the AES layout in Fig. 11.3, a set of masks are generated. The masks decompose the layout into a set of (x, y) planes, which can be vertically aligned to define the transistors and wiring layers. Layout-inserted HT are characterized as changes in one or more of the masks used in photolithography process to create physical instances of the IC. The multitude of overlapping wires and the tightly packed form of the transistors define a complex structure which represents the haystack in the "needle-in-a-haystack" paradigm. Adversaries are free to add or change very small regions in the masks, which can affect connectivity relationships between a small set of existing std. cells, or new std. cells can be added. The latter is possible using "white space," i.e., areas in the lowest layers of the layout that contain non-functional filler cells or cells implementing decoupling capacitors.

11.3.1 Layout-Oriented HT Detection Methods

HT detection methods which are designed to detect malicious modifications to the IC layout fall into three fundamental categories [54]:

- Nondestructive logic-based testing methods
- Nondestructive side-channel-based testing methods
- Destructive physical inspection techniques

11.3.1.1 Nondestructive Logic-Based HT Detection Methods

Logic-based methods derive test vectors that attempt to activate the HT [3–5, 14, 15, 48, 49]. Unlike manufacturing tests which activate and propagate faults on each node individually within the fabricated IC, HT activation is akin to multiple fault activation, which is rarely practiced in manufacturing test because of the high time complexity for ATPG and high cost of applying very large numbers of vectors. Also, unlike manufacturing defects which tend to distribute randomly across circuit nodes, the adversary chooses a stealthy location for the HT, i.e., he/she inserts the HT on circuit nodes that are difficult to control or observe [47, 61]. Unfortunately, the task of generating test vectors that provide coverage of all possible states for these nodes is orders of magnitude more difficult than it is for manufacturing defects, and, therefore, achieving high levels of HT coverage is difficult or impossible given limited resources and existing manufacturing test cost constraints. The authors of [3–5, 14, 15, 48, 49] present alternative test generation strategies that are optimized to deal with these challenges, either alone or in combination with design modifications and side-channel-based testing approaches, as detailed in other chapters of this text.

11.3.1.2   Side-Channel Analysis Approaches

Side channels refer to access and measurement techniques that bypass the designer intended input–output mechanisms, e.g., the digital I/O pins of an IC. Side channels, as the name implies, refer to auxiliary electrical and/or electromagnetic (EM) access mechanisms, such as the VDD and GND (power supply) pins or the top-layer metal connections in the physical layout of the IC. Side-channel attacks utilize these auxiliary electrical paths to introduce signals, usually in an attempt to create a fault while the IC is operational [26], or to measure signals, in an attempt to extract private internal information [28].

Side channels can also be leveraged by the trusted authority to obtain information regarding the integrity of the IC. For example, leakage current (IDDQ) and transient current (IDDT) measurements have been widely used to detect manufacturing defects [1, 2, 10, 43]. Moreover, the trusted authority can also introduce on-chip design-for-testability (DFT) [26] and other types of specialized instruments [25, 30] which allow access to additional side channels that are not directly accessible using auxiliary channels to the IC. DFT components are designed to improve visibility of the internal and localized behavior of the IC and include mechanisms to measure path delays, localized transients, and temperature profiles. DFT added by the trusted authority can also be leveraged by adversaries as "backdoor" access mechanisms to internal secrets, e.g., encryption keys, so security features such as fuses must be included to disable DFT after the IC is fabricated.

Path delay measurements, if measured at high resolutions, can also serve this role. Unlike IDDx measurements which provide a large-area regional observation, path delays are influenced by only those components that interact with the wires and gates along the sensitized path (defined as a path that propagates a logic signal transition). Therefore, path delay measurements can potentially be used to define a high-resolution HT detection methodology. Unfortunately, path delays are also affected by variations which occur in fabrication processing conditions, commonly referred to as process variations. Path delay variations caused by process variation effects are unavoidable and must be distinguished from delay variations introduced by an HT. Failure to do so is costly both in terms of the time and effort involved in verifying false alarms and, worse, in HT escapes that leave fielded systems vulnerable to attack. Subsequent sections of this chapter investigate both the benefits and challenges of using path delays as a HT detection method.

11.3.1.3 Destructive Physical Inspection-Oriented Methods

A third tactic to determining whether a chip is free of malicious inclusions is to apply destructive delayering and imaging techniques. Companies such as TechInsights (http://techinsights.com/) and Analytical Solutions (https://sstp.org/companies/analytical-solutions-inc) provide services that reverse engineer the physical characteristics of a chip to design data such as a schematic, which can then be inspected to identify IP infringements or HT circuitry. Failure analysis techniques, including scanning optical microscopy (SOM), scanning electron microscopy (SEM), picosecond imaging circuit analysis (PICA), voltage contrast imaging (VCI), light-induced voltage alteration (LIVA), charge-induced voltage alteration (CIVA), are used as needed in the reverse-engineering process [50, 54]. The primary disadvantage of these methods is their high cost and long processing times. Moreover, many destroy the chip and, therefore, cannot be used to validate chips for field use.

11.4 Fundamentals of Delay-Based HT Detection Methods

This section introduces the three fundamental technical domains that need to be considered by path-delay-based methodologies: (1) the test vector generation strategy, (2) the technique employed for measuring path delays, and (3) the statistical detection method for distinguishing between process variation effects and HT anomalies. A commercially viable HT detection method must address each of these in a cost-effective manner. We investigate the challenges associated with each of these domains and describe proposed solutions in this section. Many of the methods surveyed in Sect. 11.5 address only a subset of these technical domains and therefore must be combined with other techniques to be fully operational in practice.

11.4.1 Path Delay Measurement Schemes and Other Concepts

When technology scaling entered the deep submicron era circa 2000, higher frequency operation, within-die variations, coupling, modeling challenges, and power supply noise drove the IC design and test community to more sophisticated statistical modeling approaches for IC development and test [35, 40]. This era also renewed interest in delay fault models [38], namely, transition fault, gate delay fault, and path delay fault models, which were introduced earlier in previous works [11, 34, 39, 51]. Although it became apparent that delay fault testing was needed to keep defect levels low, workarounds were developed to allow the two-vector sequences which define a delay fault test (described below) to be applied. The workarounds became known as launch-on-shift (LOS) and launch-on-capture (LOC). LOS and LOC allow two-vector delay tests to be applied while minimizing the amount of additional on-chip logic needed to support this type of manufacturing test.

Unfortunately, LOS and LOC delay test mechanisms also create constraints on the form of the two-vector sequences, i.e., they do not allow the two vectors that define a sequence to be independently specified. These constraints reduce the level of fault coverage that can be attained for delay defects. More elaborate design-for-testability (DFT) structures that do allow both vectors of the sequence to be independently specified have been proposed [43] but are difficult to justify because of their negative impact on area and performance, and the fact that they would only be used during manufacturing test. These constraints continue to hold for modern day SoCs. However, increasing awareness of hardware trust concerns may provide the impetus for a paradigm shift which would justify additional on chip support, particularly given the significant security and trust benefits associated with path delay testing, as we discuss in the following.

11.4.1.1   Path Delay Testing Defined

Path delay tests are defined as a two-vector sequence $V1$ *and* $V2$, with the initialization vector V1 applied to the inputs of a circuit at time t0. The circuit is allowed to stabilize under V1. At time t1, vector V2 is applied, and the outputs are sampled at time t2. The Clk signal is used to drive both the launch flip-flops (FFs), which apply V1 and V2 to the combinational block inputs, and the capture FFs which sample the new functional values produced by V2. The time interval $(t2 - t1)$ is referred to as the launch-capture interval (LCI) and is typically set to the operational clock period for the chip. Figure 11.4 shows the standard form of a path delay test. Note that the standard form places no constraints on the values used for V1 and V2.

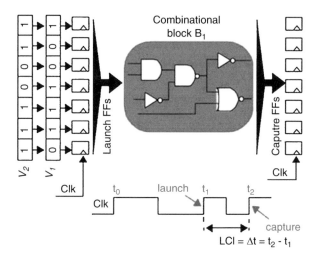

Fig. 11.4 Standard form of path delay test

Fig. 11.5 Actual form uses scan flip-flops

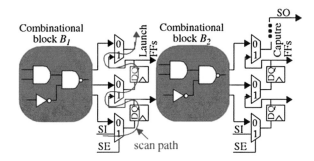

Unfortunately, external, off-chip access to the launch and capture FFs which connect to the combinational blocks within an IC is not possible. Figure 11.5 shows a typical configuration with several cascaded combinational blocks B1 and B2, with interleaved FFs. The manufacturing test community introduced a design-for-testability (DFT) feature called scan to address the difficulty of applying manufacturing tests to embedded combinational blocks [43]. Scan provides a second, serial path through all (or most) of the FFs in the IC. The second path is commonly implemented by adding a 2-to-1 MUX before the input of every FF (as shown in the figure). A scan-enable (SE) control signal is added as an I/O pin on the chip to allow test engineers to enable the serial path and to shift in a sequence of 0s and 1s using a second, I/O pin referred to as scan-in (SI). The scan path allows the internal FFs to be configured with test data that is designed to maximize fault coverage. Once a test vector is scanned into the chip, Clk is used to apply a launch-capture test, which captures the functional outputs of the blocks Bi in the capture FFs. A second scan operation allows those values to be read out using a pin called scan-out (SO).

The scan architecture shown in Fig. 11.5 allows only a single vector V1 to be applied. Manufacturing tests that target defects which prevent circuit nodes from switching (called stuck-at faults) can be applied directly using scan because the process involves applying a fixed set of values to the combinational block inputs (represented by V1) and determining if the outputs possess the correct functional values. Stuck-at fault testing is referred to as a DC test because no timing requirements exist, i.e., delays are irrelevant. As discussed at the beginning of this section, the deep submicron era brought with it more occurrences of timing related failures and the need to apply delay tests. The two-vector requirement for delay testing can be solved in two ways as discussed earlier. Launch-on-shift (LOS) derives V2 by shifting the scanned in vector V1 by 1 bit position using the scan chain. Launch-on-capture (LOC) derives V2 from the outputs of the previous combinational block, shown as B1 in Fig. 11.5 for testing paths in B2. In both cases, it is not possible to choose V2 arbitrarily as shown for the standard form in Fig. 11.4. It is important to recognize that these constraints exist (they are often ignored) and that the effectiveness of deriving delay tests for detecting HT will be negatively impacted by them.

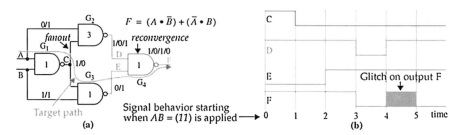

Fig. 11.6 (**a**) Reconvergent fanout and (**b**) circuit hazards

Another issue that is often ignored deals with obtaining accurate timing infor-mation for paths. The timing diagram shown in Fig. 11.4 shows that it should be possible to set the **launch-capture interval (LCI)**, i.e., the interval of time between the application of V2 at t1 and the capture event at t2, to any arbitrary value. Unfortunately, this is not the case. The external tester (ATE) driving the clock pin on the chip is limited in how close consecutive edges of Clk can be placed. Moreover, most applications of delay tests for manufacturing defects only need to determine if the chip runs at the operational clock frequency. Consequently, the LCI is typically fixed for all tests, and only upper bounds on the delays of paths within the chip can be obtained. Therefore, HT detection methods that require picosecond resolutions for individual path delays will require alternative clocking strategies and/or additional DFT components to be incorporated on the IC, which are described below.

A last important issue regarding path delay testing is related to circuit hazards. Combinational logic blocks often possess instances of reconvergent fanout. A simple example is shown in Fig. 11.6a for a NAND gate implementation of the XOR function. The integers inside the NAND gates represent one possible assignment of gate delays. The test sequence AB D f01,11g is designed to test the highlighted path but in fact propagates logic transitions along both branches of the fanout point C. The timing diagram shown on the right side of Fig. 11.6b identifies a "glitch" on the output F that is created by differences in the relative delays of these two paths.

Although this test is classified by the manufacturing test community as robust, the glitch introduces uncertainty for the security community in cases where the precise delay of the highlighted path is needed. The three transitions that occur on F each represent the delay of a subpath in the circuit, with the first, leftmost edge in this case corresponding to the highlighted path. Although subpath information might prove useful in providing additional HT coverage, process variations render this information challenging to leverage because of the difficulty associated with deciding which edge corresponds to which subpath. In other words, the same test applied to a different chip with different assignments of delays to the NAND gates may reorder the edges or may in fact result in only single transition, i.e., the glitch disappears altogether. All major synthesis tools are oblivious to hazards, making them very common in synthesized implementations of functional units. Special logic synthesis algorithms are needed to construct circuits that are hazard-free, but hazard-free implementations usually have large area overheads and therefore are rarely used. Unfortunately, hazards are largely ignored in many proposed HT test generation strategies even though they can invalidate tests and raise false alarms.

11.4.1.2 Similarities and Distinctions of Delay Test for Manufacturing Defects and HT

Unlike logic-based testing, the goals of testing for defects and testing for HT using path delay tests are very similar. Path delay tests for defects are designed to determine if an imperfection introduced during the fabrication process causes a signal propagating along a path to emerge later than designed. Similarly, path delay tests for HT are designed to determine if an adversary has added fanout to logic gate inputs and outputs, i.e., additional wires that monitor the state of the IC (trigger) or inserted additional gates in series with the original design as a means of modifying its function (payload). Both of these scenarios also cause the delay of paths to increase.

The main distinguishing characteristic between defects and HT relates to false positives. False positives are situations in which a test for an HT indicates it is present when in fact it is not. This issue is less relevant for defects and can be minimized using modern ATPG tool flows. False positives can occur for HT when the detection method does not adequately account for normal delay variations introduced by process variations. The cost associated with false-positive detection decisions is very different for defects and HT. A false positive in manufacturing test results in a defect-free chip is being falsely discarded, while a false-positive HT detection can initiate a very expensive and time-consuming reverse-engineering process of the IC.

False negatives, on the other hand, need to be handled by both manufacturing defect and HT testing communities. False negatives are situations in which a defect or HT exists, and it is not detected by the applied tests. False negatives can occur in either application either because the measurement technique does not provide sufficient resolution or because the applied tests do not provide adequate coverage. The cost associated with false negatives can be high in either case, resulting in system failure once the IC is installed in a customer application.

11.4.1.3 High-Resolution Path Delay Measurement Techniques

Delay-locked loop (DLL), phase-locked loops (PLLs), and digital clock managers (DCM) are on-chip IP blocks responsible for maintaining phase alignment with external oscillators and for creating multiple internal clocks at different frequencies and with specific phase shifts. They can be used to create the Clk signal shown in Fig. 11.4 for path delay testing. Although automatic test equipment (ATE) can be used to carry out path delay testing, on-chip clock and phase shift mechanisms

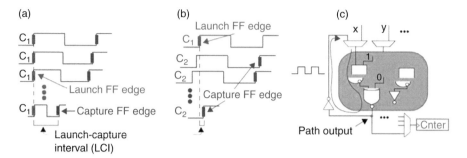

Fig. 11.7 Path delay measurement techniques. (**a**) Single-clock scheme. (**b**) Dual-clock scheme. (**c**) RO scheme

generally provide higher accuracy and resolution because off-chip parasitic resistor-inductor-capacitor (RLC) components are eliminated and noise sources are reduced. Many of the HT detection techniques described in subsequent sections depend on high-resolution timing measurements, making on-chip techniques better suited.

Figure 11.7 shows three examples of measurement techniques that can be used to provide fine-grained timing resolution. The first, called single-clock scheme (or clock sweeping), requires repeated application of a two-vector sequence (Fig. 11.7a). On each iteration, the frequency of C1 is increased, which moves the launch and capture edges, i.e., the LCI, of the Clk signal closer together. The process is halted as soon as a condition is met or violated, which is usually related to whether the capture FF successfully captures the functional value produced by vector V2 (see Fig. 11.4). An estimate of the path delay is computed as 1/frequencyfinal where frequencyfinal is the stop point frequency. Although this scheme requires the fewest resources, i.e., only one clock tree is included on the chip, it lower bounds the length of the path that can be measured.

For example, short paths would require a very high-frequency clock, which creates undesirable secondary effects, e.g., power supply noise, that make it difficult to obtain accurate timing measurements. Single-clock schemes which use an externally generated (ATE) clock constrain the minimum path length even further.

The second, called dual-clock, scheme (or clock strobing), also requires repeated application of the two-vector sequence [20, 31]. On each iteration, the phase of the capture clock C2 is decremented by a small _t relative to C1 as shown in Fig. 11.7b. The additional overhead introduced by the second clock tree is offset by the benefit of being able to time a path of any length.

This is possible because the two clock networks are independent and modern clock manager IP designs are capable of allowing the time base of C2 to be very precisely shifted. Moreover, the power supply noise issue mentioned above is also mitigated because only two clock edges are required to carry out a launch-capture delay test instead of three.

The third timing mechanism, referred to as the RO scheme, is shown in Fig. 11.7c [50] (a similar scheme called Path RO is proposed earlier by the authors of [55] but

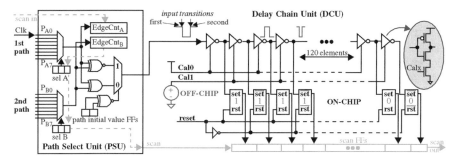

Fig. 11.8 Time-to-digital converter (TDC)

for application to design for manufacturability), it adds the components shown in magenta to the design. Paths in the circuit are timed by creating a ring oscillator (RO) configuration where the output of a path is connected back to the input of the path using a MUX (and optionally a NOT gate as shown). A timing measurement is performed by enabling the MUX connection and then allowing the path to "ring" for a specific time interval. A counter (Cnter) is used to record the number of oscillations. This is accomplished by tieing the output signal from the path to the clock input of the counter. The actual path delay is obtained by dividing the time interval by the counter value (note, the NOT gate and MUX add two gate delays to the delay of the actual path). No launch-capture event is required in this scheme. Therefore, the clock noise associated with high-frequency clocks in the single-clock scheme is eliminated. The main drawback is related to the limited number of paths that can be timed in this fashion. For example, paths that have hazards as discussed in reference to Fig. 11.6 produce artifacts in the count values. As discussed, hazards are very common in combinational logic circuits, and therefore, they will negatively impact HT cover-age.

A fourth alternative, called a time-to-digital converter (TDC), is shown in Fig. 11.8 [22, 25, 30]. Similar to the RO scheme, it eliminates clock strobing and, therefore, is able to obtain path delay measurements that better represent mission mode path delays. The TDC is an example of a "flash" converter, which is a class of converters that digitize path delays very quickly. The path select unit shown on the left is responsible for selecting a pair of paths, one of which can be the clock signal. The delay chain unit shown on the right is responsible for creating a digital representation of the relative difference between the delays of the two input paths, PAx and PBx.

The arrival of a rising or falling transition on one path creates the first edge in the delay chain (labeled first in the figure), while a transition on the second path generates the trailing edge (labeled second). The width of the initial (leftmost) pulse shown in red represents the delay difference between the two signals being timed. The pulse propagates along the delay chain as shown by the annotations along the top of the figure. The inverters in the delay chain include an additional series-inserted NFET transistor as shown by the callout on the far right. An analog control signal labeled Calx is used to control the pull-down strength of the inverter, with higher gate voltages allowing faster operation. The inverter chain is configured with two such control signals, Cal0 and Cal1.

The combination of the two allows independent control over the propagation speed of the first and second edges. A calibration process is carried out in advance of making delay measurements to determine the best values of Cal0 and Cal1. These analog control signals are set to allow the worst-case (widest) pulse to propagate through most of the inverters before "disappearing." The pulse disappears when the second edge catches up to the first edge. The calibration process is described later in Sect. 11.5. The output of the inverters in the delay chain also each connects to a "set-reset" latch. The presence of a negative pulse (for odd inverters) or positive pulse (for even inverters) changes the latch value from 0 to 1. A digital thermometer code (TC), i.e., a sequence of 1s followed by zero or more 0s, is produced in the sequence of latches after a test completes. The TC can be converted into a discretized delay value (if needed) using pulse width information applied during the calibration process. In addition to being very fast (less than 100 ns per measurement), the TDC is also resilient to some types of circuit hazards. For example, a series of pulses can be introduced by circuit hazards but only the widest one determines the TC value (shorter pulses die out earlier in the delay chain). The EdgeCnt components in the path select unit can be used to decide when hazards are present.

A fifth scheme, called REBEL in [10], also uses a delay chain to obtain timing information. REBEL is a light-weight embedded test structure that combines the delay chain component of the TDC (without the pulse shrinking characteristic) with the clock strobing technique referred to in Fig. 11.7. A significant benefit of REBEL over the TDC is complete resilience to circuit hazards. In fact, REBEL is able to provide timing information regarding each of the edges associated with hazards in a single launch-capture test. As indicated earlier, the edges produced by circuit hazards each represent the delay of some internal segment in the functional unit. Although process variations add uncertainty and diminish their usefulness, as discussed above, the ability to instantly have knowledge of their presence adds robustness to the delay measurement process and can help reduce the likelihood of false-negative HT detection decisions.

11.4.2   Dealing with Process Variations

A significant benefit of techniques designed to detect HT in fabricated chips is the availability of a golden model of the IC. The assumption made by most of the techniques described in Sect. 11.5 is that the HT is introduced by changing the layout, via mask manipulation or through other fabrication process-related steps. Therefore, all design data prior to the mask and chip fabrication steps, e.g., HDL, schematic, and even the geometric layout data itself, are considered trusted. A golden model, and simulation data derived from it, provides a trusted reference to which hardware data can be compared. Path delay methods attempt to identify anomalies in the hardware data that cannot be explained by the golden model.

The most significant challenge associated with detecting HT with path delay testing is distinguishing between changes in delay introduced by an HT and those introduced by process variation effects. Failing to distinguish between these two types of delay variations leads to false-positive and false-negative HT detection decisions. The former declares an HT is present when it is not, while the latter fails to detect the presence of an actual HT. Minimizing false-positive and false-negative rates is a critical design parameter of an HT detection technique.

There are three basic approaches for dealing with process variation effects in HT methods. The first method, called **GoldenChip-based**, creates simulation models or uses HT-free chips, respectively, to characterize the HT-free space. The second, called PCM-based, uses data from process control monitors (PCM) to "tune" the boundaries of the HT-free space derived from golden models using chip-measured test structure data. The third, called Chip-Centric, creates a nominal simulation model and calibrates and averages path delays to the nominal model (or data from HT-free chips). All approaches create a bounded HTfree space that represents normal variations in path delays introduced by process variations and/or measurement noise. Data collected from the test chips is compared with this bounded HT-free space. Data points that fall outside the boundaries are called outliers, e.g., are path delays that exceed the limits defined by the HT-free space. Chips that produce outlier data points are considered HT candidates.

The three approaches are graphically portrayed in Fig. 11.9 and are described in more detail throughout Sect. 11.5 as needed. The 2-D shapes labeled "Simulations with process variations modeled" and "Delay variations across chip population" can in fact be multidimensional, with each dimension representing one path delay or one of multiple features extracted from the set of path delays using statistical techniques such as principle component analysis (PCA).

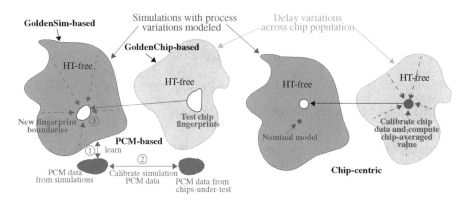

Fig. 11.9 Mechanisms to account for process variations using a golden model approach for HT detection

GoldenChip-based and GoldenSim-based techniques train a classifier using HT-free data from chips or simulations, respectively. GoldenChip-based methods measure delays from HT-free chips, which are then destructively validated to be HT-free using techniques from Sect. 11.3.1.3. GoldenSim-based methods typically use data from Spice-level simulations of a resistor-capacitor-transistor (RC-transistor) model of the golden design. Both techniques can be expensive in terms of reverse-engineering effort, model development, and simulation time. Delayering technologies utilized for GoldenChip-based methods can take weeks or months. For GoldenSim-based methods, CAD tools such as Mentor Graphics Calibre must first be used to create the RC-transistor models of the layout using complex process models obtained from the foundry in which the chips are fabricated. The modeling files can be very large, e.g., 100s of MB, even for relatively small designs on order of 20,000 gates, and simulation times can easily extend to weeks and months when performing transient simulations with only a couple 100 input vectors. The effort required to construct and/or confirm the HT-free boundaries using either technique is very large and is often underreported.

Of even greater concern for GoldenSim-based techniques is the level of mismatch that can exist between the simulation results and the hardware. Foundry models in advanced technologies have become very complex, providing the user with a variety of statistical evaluation methodologies including fixed corners and Monte Carlo options. Fixed corner models are provided to enable the user to predict worst-case and best-case performance of the chip by modeling the range of global process shifts that can occur over time. Unfortunately, this typically expands the HT-free space beyond what is required to represent the behavior of the chips under test. The expansion leads to a decrease in the sensitivity of HT methods and increases the level of mismatch between simulation and hardware data. Moreover, foundry models typically provide limited capabilities for modeling within-die variation effects, making it difficult to predict the uncertainties related to specific hardware path delays. These modeling and simulation challenges are compounded by measurement noise that occurs during chip testing and by non-zero jitter and drift tolerances introduced by the tester during the generation and delivery of high frequency clocks. Taken together, these issues work to increase in the possibility of false-positive and false-negative HT detection decisions.

11.4.3   Test Vector Generation Strategies

The last issue deals with an important distinction that exists between fault models used in manufacturing test and those required for detecting HT. The manufacturing test community developed several fault models, including transition delay faults and path delay faults, for dealing with timing problems resulting from a wide variety of defect mechanisms. For example, the transition delay fault (TDF) model assumes that defects occur on individual nodes in the circuit and that they manifest as slow-to-rise and slow-to-fall signal behaviors at those nodes. The path delay fault (PDF) model, on the other hand, makes no such assumptions, i.e., it accounts for defects which may in fact be distributed across one or more logic gates and wires that define the paths, and as a result, the PDF model provides more complete information about the integrity of the tested chip.

Unfortunately, obtaining 100% PDF coverage requires all (or a large fractions) of the paths in a chip to be tested. For even moderately sized circuits, the costs associated with the generation and application of a complete PDF testset is prohibitive. This is true because the number of paths can be exponentially related to the number of inputs to the chip (or functional unit). Therefore, most chip companies generate and apply TDF vectors instead because the number of such tests is linear to the number of circuit nodes in the design. Fortunately, for the security and trust community, the TDF model is a better match to the types of malicious modifications an adversary is likely to make to the layout. The node-oriented TDF model used for defects is leveraged by a large fraction of the proposed HT detection techniques described in Sect. 11.5.

There are two important points to consider with regard to test generation for HT detection. The first relates to the options that are available when the TDF model is used. Although far fewer tests are required under the TDF model to obtain high levels of HT coverage (when compared to the PDF model), there are typically many choices for the path that is sensitized through each of the nodes. A variety of techniques are proposed by authors of published work including random vectors, an incremental coverage strategy driven by the sequence of vectors generated so far, traditional TDF vectors, or, in some cases, the test generation strategy which is left unspecified. Others leverage the TDF model and direct ATPG to target the shortest paths through the node because the additional delay added by the HT (via fanout load or gate insertion) has a larger fractional impact on the path delay. The traditional TDF model for defects, on the other hand, typically target the longest paths as a mechanism to ensure that at least this subset of tested paths meet the timing constraints.

The length of the path relates to the second important point regarding test generation. Automatic test equipment is outfitted for manufacturing test, which is focused on testing the longest paths. For test cost reasons, it is common that only one clock frequency is used to apply TDF tests to the chip because the primary goal of manufacturing test is to ensure that the delays of all tested paths are less than the upper bound defined by the clock period. The most sensitive tests for defects therefore are those that test the longest paths. This is true because the longest paths minimize the slack, i.e., the difference between the clock period and the delay of the tested path.

Many believe that these manufacturing test constraints for defects are not sufficient for providing high levels of HT coverage. This is reflected in the proposed use of clock sweeping, clock strobing, and other on-chip embedded test structures for obtaining precise measurements of path delays. In other words, the slacks inherent in tests for defects provide too many opportunities for adversaries to "hide" the additional delay of the HT in the slack, and, therefore, a paradigm shift is required regarding the manner in which delay testing is carried out on the test floor.

Clock sweeping and clock strobing are expensive in terms of test time, and HT detection techniques which use these clocking strategies need to account for the higher levels of clock noise associated with high-frequency clocks and invalidations introduced by circuit hazards. It remains to be seen how the economic trade-offs of delay-based HT detection schemes will play out.

11.5   HT Detection Methods Based Path Delay Analysis

This section is dedicated to describing a selected subset of the proposed HT detection strategies that have been proposed over the last decade. Our goal is to describe methods that offer some unique perspective, and therefore, this exposition does not provide an exhaustive survey of every published paper on the topic. The choice to include a description of a published work was based on whether it promoted the state of the art in at least one of the three technical domains described earlier, including the path measurement technique, the statistical method used to distinguish between HT anomalies and process variation effects, and the test vector generation strategy. The techniques are presented chronologically instead of by technical domain. The latter organization presented challenges because many techniques propose solutions to more than one domain.

11.5.1   Early HT Detection Techniques and On-Chip Measurement Methods

The first works on using path delays for HT detection are described in [24] and [32]. The primary focus of each paper is on only one of the technical domains, in particular [24] on a statistical method for distinguishing between process variations effects and HT and [32] on a high-resolution on-chip measurement technique.

In [24] the HT detection method assumes that high-resolution path delay measurements are available, i.e., no measurement strategy is proposed. Although not explicitly stated, the test vector generation strategy appears to be based on the standard transition delay fault model. They base their detection method on the GoldenChip-based model described in Sect. 11.4.2. A multivariate statistical technique is used to extract distinguishing features from the full set of path delays. HT-free chips are used to construct the HT-free boundaries, which they refer to as a fingerprint. The fingerprints define the boundaries of the shape labeled "Delay variations across chip population" shown on the left side of Fig. 11.9. HT are detected by comparing the delay fingerprints measured from the untrusted test chips with the boundaries defined by the HT-free fingerprints.

They demonstrate their technique using simulations in which ATPG-derived two vector sequences are applied to a DES functional unit. Principle component analysis

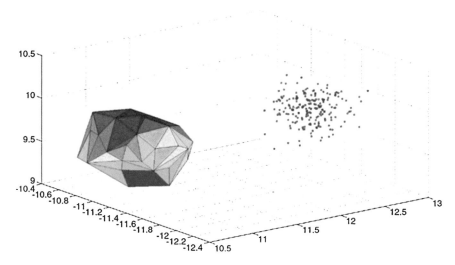

Fig. 11.10 Convex hull characterization showing detection of an HT [24]

(PCA) is used to extract distinguishing features from a set of 10,432 simulated path delays as a means of reducing the HT-free space to a 3-D structure. A statistical technique based on a convex hull characterization of the HT-space is used to define the boundaries for each of the 64 outputs of DES. Four HT are inserted into another set of models, with three representing explicit payload HT and one representing an implicit payload HT. The explicit payload HT inserts one or more additional gates in series with paths in the HT-free design, while the implicit payload HT is represented as a simple counter with no ability to change the functional characteristics of DES. They show that the explicit payload HT are easily detected (see Fig. 11.10), while the implicit payload HT is only detected approx. 36% of the time.

A high-resolution on-chip path delay measurement technique is proposed in [32], which is extended in [44] to include a GoldenSim-based HT detection strategy. The measurement technique is based on the dual-clock scheme described in reference to Fig. 11.7. A set of shadow registers are added to each of the outputs from the combinational components of the design, next to the capture FFs or destination registers as shown in Fig. 11.11a. The second clock of the dual-clock scheme, CLK2, is used to drive the clock inputs of the shadow registers. CLK2 is generated as a "fine-phase-shift" adjusted version of CLK1 using a DCM on the FPGA.

The process of measuring the path delay of the combination path from Fig. 11.11 begins by setting the phase shift of CLK2 to a small negative value, on order of 10–100 ps (see Fig. 11.7). A two-vector sequence is applied to the source registers using a launch-capture test. The comparator, also added to the design, is used to determine if the captured values in the destination and shadow register are the same or different. If they are the same, which is the case when the clock strobe operation begins, the negative phase shift difference between CLK1 and CLK2 is increased, and the same two-vector sequence is applied. This process is repeated

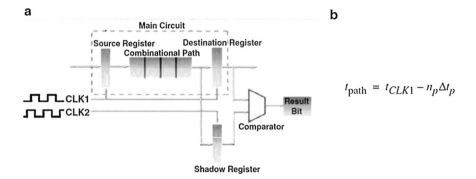

Fig. 11.11 (a) On-chip path timing architecture proposed in [32] and (b) equation giving actual

until the comparator indicates the values are different. The actual delay of the path is computed by multiplying the number of phase shifts, np, by the _tp provided by each phase shift increment. Subtracting this value from the CLK1 period yields an estimate of the path delay, tpath, as given by the equation in Fig. 11.11b.

The extended work in [44] investigates the detection capabilities of a GoldenSim-based technique on an 8-bit Braun multiplier functional unit. HT are modeled as series-inserted two-inverter chains. The proposed method derives a path delay distribution using simulation data but is constrained by the measurement resolution provided by the timing technique. Process variations are modeled by varying transistor threshold voltage, Vth, and transistor channel length, Leff, in simulations with and without HT. Data from these simulations is used to define the boundaries of the shape labeled "Simulations with process variations modeled" shown on the left side of Fig. 11.9. The amount of skew in the mean of the distributions is used as the detection criteria. The results using four inserted HT show that three can be detected and the last one is detectable on some outputs but not others.

11.5.2   Ring Oscillator-Based HT Detection Approaches

A distributed set of ring oscillators (RO) is proposed in [63] as a means of detecting HT. The array of ROs is distributed uniformly across the (x,y) space of the functional unit as shown in Fig. 11.4. The detection criteria are based on HT power consumption. If an HT is present, the test stimulus applied to the functional unit may cause at least some gates within the HT to switch (referred to as partial activation in [54]), and the HT will necessarily consume power. The additional HT power consumption creates localized voltage drops on the supply rail (VDD) that can be detected by comparing the delay of a nearby RO with that of an HT-free chip or simulation. The delay variations introduced by the HT-switching-induced voltage drops in the RO are integrated by the RO over time and are reflected in a counter

Fig. 11.12 RO distribution architecture proposed in [63] for detecting HT switching activity

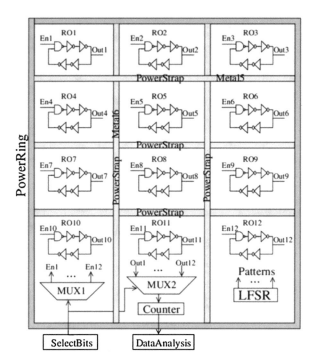

value. A counter is connected to the RO using MUX2 as shown along the bottom of Fig. 11.12. MUX1 is used to select and enable one RO at a time as a random, LFSR-based test vector sequence is applied to the inputs of the functional unit. This process is repeated for the n ROs (12 in figure) with the set of counts representing the signature for the chip.

Process parameter variations are accounted for by collecting data from a large number of HT-free chips (GoldenChip-based model), and a statistical analysis is applied to the signature using principle component analysis (PCA) and correlation analysis. One of the techniques, called advanced outlier analysis, analyzes data obtained from pairings of ROs as a means of detecting regional power droop anomalies. The results shown in Fig. 11.13 plot the RO pairings that show the best detection capability for each of the six HT inserted into the design. The points within each graph represent experimental results with process variations derived from FPGA experiments. The separation of the red HT-inserted and blue HT-free points illustrates that nearly all are detected in every FPGA.

The authors of [45] propose a design-for-trust (DFTr) technique designed to detect HT by creating ROs from the functional paths in a design (a related design-for-manufacturability scheme was proposed earlier in [55] for measuring critical path delays). They propose an algorithm that selects paths with the maximum number of "unsecured gates," i.e., gates that have not already been included in other ROs, i.e., a method characterized as an incremental coverage-driven strategy. For each selected path, a MUX, a control signal, and optionally an inverter are added to

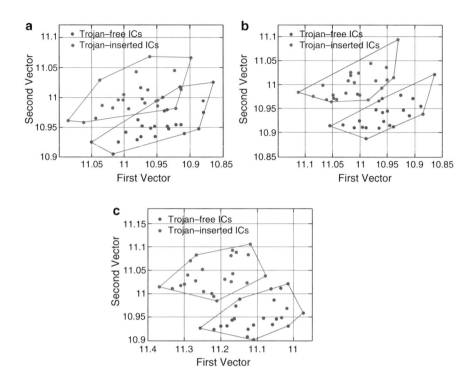

Fig. 11.13 Subset of results from [63] using proposed advanced outlier analysis method on FPGAs. (**a**) T7. (**b**) T8. (**c**) T9

the design to complete the ring. Automatic test pattern generation (ATPG) is then used to generate input patterns that place nondominant values on the off-path inputs of gates sensitized by the ring. Nondominant value refers to gate input values that do not determine the gate's output value by themselves, e.g., a "1" is the nondominant value for an AND gate. Off-path inputs refer to side inputs of gates in the RO that are not on the sensitized path of the RO. These conditions ensure that the RO will "ring" when enabled by the control signal.

Figure 11.14 shows the ISCAS-85 benchmark circuit C17 configured with a set of ROs. Simulation experiments with process variation effects modeled are carried out to determine the golden frequencies, Fgolden (GoldenSim-based model). HT detection is carried out by comparing Fmeasured obtained from each of the untrusted test chips with Fgolden. The authors implemented their technique on a Xilinx Spartan 3 FPGA using six ISCAS-85 benchmark circuits. The number of ROs required to attain a specific level of HT coverage is given in Fig. 11.15a. As is typical of test generation for manufacturing defects, coverage per RO drops dramatically for coverage targets above 90%. Test times are given in Fig. 11.15b which shows similar trends. Although the proposed technique is very promising, the authors do not address hazards that occur within circuits with reconvergent fanout.

Fig. 11.14 ISCAS-85
benchmark C17 configured
with a set of Ros using the
DFTs method proposed in
[45]

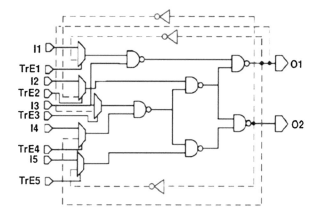

Fig. 11.15 (**a**) Number of
ROs required as a function of
coverage for six ISCAS-85
benchmark circuits, and (**b**)
test time required to obtain
$F_{measured}$ for each chip [45]

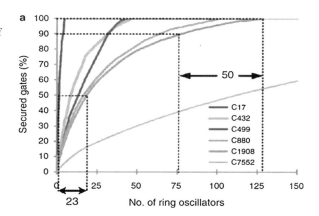

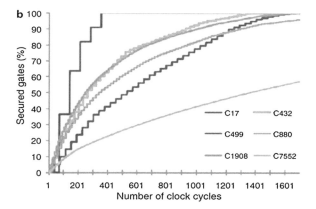

11.5.3 Lightweight On-Chip Path Timing Techniques for HT Detection

Two on-chip delay measurement techniques, called TDC and REBEL, are proposed in [10]. An HT detection method which leverages REBEL [29] is described in the following. A second HT detection method which uses the TDC [22] is covered in Sect. 11.5.11. REBEL is designed as an embedded test structure (ETS). ETS is an instrument that is integrated directly into functional units as a mechanism to obtain regional, high-precision in-formation about its operational characteristics. ETS must be designed to be minimally invasive and low in overhead to avoid violating power, area, and performance constraints associated with the functional unit. REBEL satisfies these ETS attributes by leveraging components in the existing scan chain architecture.

REBEL is designed to provide regional, high-resolution measurements of path delays. It also addresses the clock noise issue discussed in Sect. 11.4.1 related to using single-clock schemes to obtain timing information for short paths. The architecture of REBEL allows paths within the functional unit to be extended along a delay chain, effectively eliminating the need for high-frequency clocks. The delay chain is created using the existing capture FFs attached to the outputs of the functional unit. Figure 11.16a shows an example configuration with REBEL integrated into a pipelined functional unit. A path under test (PUT) within the functional unit is highlighted as well as the delay chain that is created through the capture FFs. Row control logic is added to the design to enable one of the path outputs to be selected as the target of the timing measurement process.

A path is timed by applying a two-vector sequence to the inputs of the functional unit. The transition along the PUT emerges at the output and propagates along the delay chain. The capture edge of the clock creates a digital snapshot of the transition by storing in the capture FFs a sequence of digital values which represent its behavior over time. Each of the snap-shots immediately reveal whether the propagating signal has more than one transition, i.e., whether a hazard is present or not. A sequence of digital snapshots is shown in Fig. 11.16b as rows labeled 120 to 180, which represent a clock sweeping sequence of LCIs as described earlier in reference to the single-clock scheme of Fig. 11.7. For each successive LCI, the propagating falling transition driving the input of the capture FF15 is given more time to propagate along the delay chain. The path tested in this example does not generate any type of hazard (is hazard-free); otherwise, one or more of the snapshots would show interleaved "1s" in the sequence of "0s." In practice, the LCI test sequence is actually applied in reverse, starting with 180, because larger LCI increases the amount of temporal information stored in the capture FFs regarding the propagating transition. The larger time window provides a better opportunity to detect hazards which can invalidate the HT test.

As proof of concept, a 90 nm chip is designed and tested which allows paths in the functional unit (in this case, an eight-function floating point unit or FPU) to be reconfigured with and without an HT. Although the experimental results

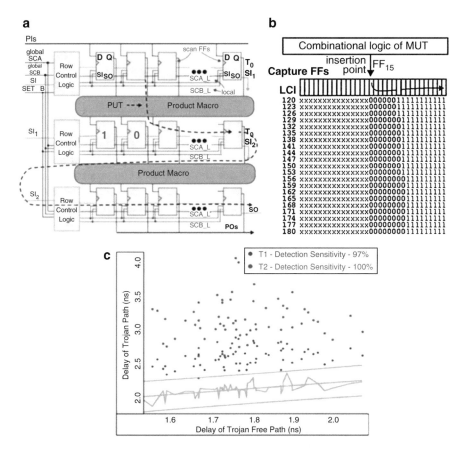

Fig. 11.16 (**a**) Integration of an embedded test structure called REBEL in a pipelined functional unit as described in [29], (**b**) sequence of digital "snapshots" stored in the REBEL delay chain with successive rows showing an increasing launch-capture interval (LCI) and (**c**) illustration of regression analysis applied to path delays measured from 62 chips for an HT-free path (*x*-axis) and a second path (*y*-axis) with and without the inclusion of HT

presented use hardware data to define the HT-free space (GoldenChip-based model), the authors also acknowledge that a GoldenSim-based approach is possible.

The statistical method proposed in [29] is based on linear regression analysis. Figure 11.16 shows a subset of the results in which delays from one HT-free path (*x*-axis) and a second path that can be configured with and without HT (*y*-axis) are compared. The HT-free space is highlighted in green and is delineated by three sigma regression bands. It is derived using data from 62 copies of the test chips configured without the HT included in the second path. The red and blue points each represent the results with the second path configured with one of two possible HT. All but one of the data points fall outside the regression limits and are therefore classified as detected. Additional results for a larger set of HT are reported in the paper.

11.5.4   Self-authentication: A Golden Model-Free HT
Detection Method

A golden model-free HT detection method is described in [33] which inserts a framework of HT detection sensors into the layout representation of a design. The sensors are designed as replicas of common sequences of logic gates (path sequences) that already exist in the design[1]. Custom CAD tools are used to decompose the timing graph of a design to identify a set of commonly occurring delay features. Delay features correspond to layout-specific patterns of gates and interconnect that share common geometries and sensitivities to process parameter variations. Similarity among features is determined by evaluating the changes in delay that occur when the path sequences are subjected to similar process conditions. Two sequences are considered similar if the changes in their delay track within a small error tolerance.

Once a set of target path sequences are identified in the design, a set of matching sensors are integrated into the layout in close proximity to the targeted path sequences. After fabrication, the delay fingerprints of the sensors and corresponding full-length paths (that contain the path sequence(s)) are measured. Data from each of the sensors is used to construct an HT-free delay range, which captures the measurement noise profile for the sensor. A similar process is carried out for each of the paths. The delay associated with other components of the full-length paths, in which the path sequences are contained, is accounted for using variation-aware expressions. A nominal simulation or static timing analysis estimate is used to determine the nominal delay of the sensor, which, in combination with the measured delays, allows the delay of the full-length paths to be predicted. Correlation analysis is used to compare the predicted and measured delay ranges for the sensors and paths. Outliers are considered anomalies introduced by HT in either the sensor, the path, or both. Figure 11.17 provides a flow diagram of the self-authentication process and shows examples of the HT insertion and detection scenarios considered [33].

The authors assume on- or off-chip delay measurement schemes such as those described in reference to Fig. 11.7 are available. The sensors act as silicon-anchor points for calibration, and therefore the proposed HT detection technique shares similarities to the process control monitor approach described in Sect. 11.4.2 and referred to as "PCM-based."

The authors apply the technique to the ISCAS-89 benchmark circuits, synthesized to layouts using a 90 nm TSMC technology. Process variations are modeled in the simulations by varying major device parameters within 10% of nominal using a Monte Carlo selection process. A multilevel hierarchical model of the layout is processed as a means of partitioning the design into regions where it is assumed that process variations are more highly correlated. Sensors are identified and designed but constrained to use no more the 15% additional area in the layout. A set of

---

[1]A self-referencing technique is also proposed earlier in [19] but is based on the correlation of transient power supply currents produced by replicas of the functional unit.

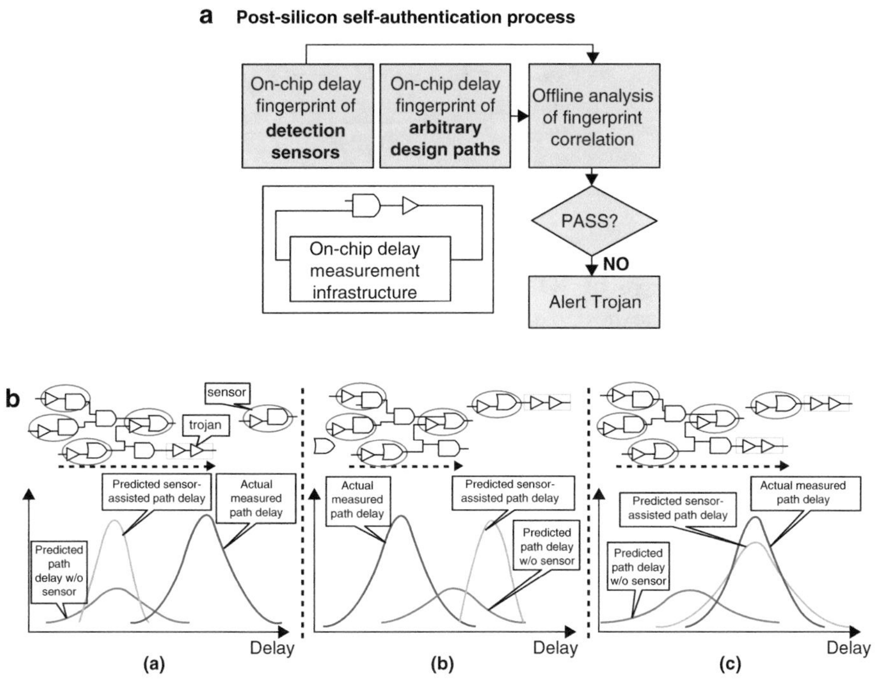

Fig. 11.17 (**a**) Self-authentication chip testing process proposed in [33] and (**b**) sensor (both simulated and measured) and measured full-length path delay range illustrations under three HT attack models. Mismatches in the overlap among the sensor distributions or low levels of correlation between sensor and path distributions is flagged as an HT detection

paths are randomly selected to serve as the HT insertion points for evaluation of the method. A set of 30 HT are inserted into each path with varying amounts of delay to determine sensitivity, and 10 K process models are created and simulated (300 K per path). HT detection rates are shown to improve from between 2 and 16% when compared to a similar method that does not leverage sensors as a sensitivity enhancing technique.

11.5.5 Linear Programming Methods and Test Point Insertion for HT Detection

Linear programming method is proposed to derive leakage, power, and delay characteristics of individual gates based on solving a system of equations, referred to as gate-level characterization (GLC) [42, 56, 57]. Chip measurements of power and delay are used in the system of equations, along with estimates of measurement errors, to derive scaling factors for the parameters associated with the logic gates in the design.

Fig. 11.18 Circuit showing reconvergent fanout [57]

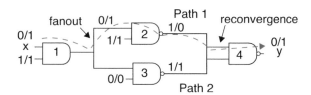

GLC is combined with a test point insertion technique in [57] as a means of improving coverage of HT. The authors propose to add FFs (test points) to components of the design that exhibit reconvergent fanout. An example of reconvergent fanout is shown in Fig. 11.18 from [57] where both the fanout and reconvergence points are identified. The task of generating path delay tests for circuits which contain complex reconvergent fanout networks is an NP-complete problem. Automatic test pattern generation (ATPG) algorithms can fail to determine two-vector sequences that are able to test the individual paths within reconvergent fanout blocks, such as those labeled "path 1" and "path 2," even when such tests exist. Although in the example it is trivial to derive test patterns that test these two paths individually (node assignments are shown that allow path 1 to be tested by itself), there are other more complex configurations which require an exhaustive search, proportional to 2n, to find suitable two-vector sequences.

The authors propose an algorithm that first identifies all paths that can be easily tested and then a set of paths in reconvergent fanout logic structures that are the best candidates for test point insertion. A SAT-based process is proposed to select input vectors that maximize the number of independent linear equations for application of GLC. A second circuit partitioning scheme based on maximum fanout-free cones is proposed in [56] to increase the number of delay access points within the circuit as a means of improving coverage further for large designs.

11.5.6   Process Calibration and Test Vector Selection for Enhancing HT Detection

A delay calibration technique is described in [12] that leverages information obtained from test structures as a means of detecting HT delay anomalies that are very small, i.e., within the margin of those introduced by process parameter variation effects in advanced technologies. The test structure measurements are used to estimate the global mean shift in delay introduced by variations in the process parameters for each chip. Based on the estimate, the mean value for the paths in the region of the embedded test structures is calibrated to eliminate the mean shift.

The process flow proposed in [12] is shown in Fig. 11.19. The first step is to extract information from the embedded test structures, such as ring oscillators, as a means of obtaining process parameter information for each chip. Test structures are added to the layout in regions close to the functional unit to be tested. This ensures that both global process variations and systematic within-die variations are accurately captured in the measurements. Path selection and vector generation are carried out such that test cost is minimized. ATPG is constrained to generate robust tests (when possible) for critical (longest) paths passing through each possible HT site, and therefore test generation is based on the traditional TDF model.

Path delays are measured using the dual-clock scheme shown in Fig. 11.7 as a means of minimizing clock noise and obtaining high-resolution measurements. The integration of silicon anchor points for calibration of process variations classifies the proposed technique as PCM-based. An estimate of the mean shift in each region of the chip is computed using test structure data and a minimum mean square (MMSE) estimator. The MMSE finds the mean that minimizes the sum of the squared differences between the test structure data and the computed mean as given by Eq. 11.1. This estimator can then be used to calibrate all the path delays for a given chip by simply subtracting the mean from each measurement.

$$\sum_{i=1}^{M} (di - \mu)^2 \tag{11.1}$$

A novel nonparametric hypothesis testing method based on a likelihood-ratio test is proposed which leverages integer linear programming (ILP) for determining the number of chips that need to be tested to achieve a specific confidence level against false-positive and false-negative HT detection decisions.

The technique is evaluated on a set of inverter chains of length 2 through 12 with and without HT insertions. HT are modeled using a minimum size inverter connected to the output of the first inverter in the chain. Monte Carlo simulations were performed using circuit models with different types of global and within-die process variations modeled, referred to as case I (global only), case II (across-chip random and systematic within-die only), and case III (local random and systematic within-die only). Figure 11.19b shows that the best results are obtained for case III which uses simulation models with only local process variations included. The proposed calibration method in this case makes this possible by eliminating case I and case II via the test structure measurements, which minimizes the _/_ statistical variation parameters as well as the number of required chips.

This technique is extended in [13] to address the best paths to target for HT detection. In contrast to their earlier work, the authors argue that the shortest paths through each HT site maximize detection sensitivity (see shortest path TDF discussion in Sect. 11.4.3). The column labeled _/_ in Fig. 11.19b expresses the impact of the HT on path delay and is the focus of the current work. Given that the adversary's goal is to minimize the impact of the HT on path delay, shorter paths are better suited to reveal these small delay variations because the (constant) delay added by the HT becomes a larger fraction of the total delay for short paths. A similar argument regarding the effect of process variations also holds. In particular, the _ of variations is approx. Proportional to the nominal delay of the path, i.e., shorter paths have smaller _. This characteristic is illustrated in Fig. 11.20 which

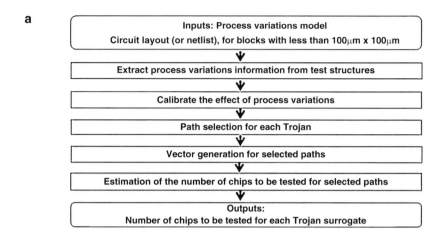

Fig. 11.19 (**a**) Process flow model proposed in [12], (**b**) simulation-derived delay statistics obtained by applying the proposed method under different types (global and within-die) process variations

shows the path distributions for a long path (top) and short path (bottom) with and without HT. The HT, represented as an "additional fanout," creates a more distinguishable shift in the short path distribution when normalized as a fraction of the total width of the distribution. In both cases, the HT adds only 8 ps to the path delay but the smaller _ corresponding to the shorter path provides a higher level of confidence in detecting the anomaly.

The authors also argue that shorter paths are more likely to be the targets of an HT insertion because longer paths, particularly critical paths, increase the chance of accidental discovery. Moreover, generating vectors for shorter paths is generally "easier" for ATPG tools to accomplish because fewer side inputs need to be "justified" (forced to specific values) in order to sensitize the path from PI to PO. The main benefit of short paths, however, according to the authors, is the reduction in the number of chips that need to be tested (see column labeled N in Fig. 11.19b). On the downside, shorter paths are harder to time, especially when using the single

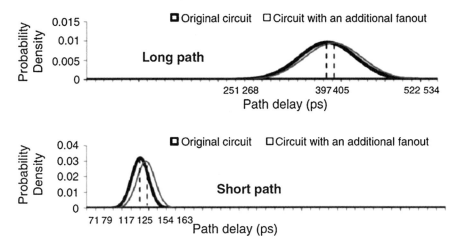

Fig. 11.20 Impact on path delay distribution for a long and short path, with short path showing larger fractional change [13]

clock scheme from Fig. 11.7, because the chip needs to be tested at much faster than at speed to obtain precise delay measurements. The dual-clock scheme provides a solution, but it also requires the addition of a second clock tree as described in [32]. An algorithm is presented that both selects the shortest path through each circuit node (each HT site) and enforces constraints on the robustness of the test to ensure the target path is in fact the path tested by the two-vector sequence. The authors present simulation results using the ISCAS-89 benchmark circuits that show a 2.1× improvement in test cost over a traditional TDF strategy. They further show that the improvement increases to 4.51× when combined with the calibration technique proposed in [12].

11.5.7 Clock Sweeping for HT Detection

The authors of [60] propose a clock sweeping method to address sensitivity issues associated with using a traditional TDF model and the path delay fault (PDF) model for detecting HT. Clock sweeping refers to the single-clock scheme referenced earlier in Fig. 11.7 in which the clock frequency is incrementally increased (by a fixed step size) and a two-vector sequence is applied repeatedly until a delay fault is detected in the capture FFs for one or more of the tested paths. The authors propose to generate tests using the TDF model described earlier and acknowledge that short paths whose delay is smaller than the maximum frequency are not testable because of the limits of ATE and clock noise. The algorithm that they propose is shown in Fig. 11.21a. It partitions TDF tests into two groups, those sensitizing long paths to be tested in the proposed HT delay technique and those sensitizing short paths to

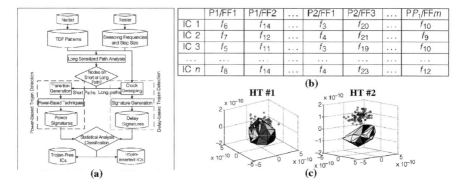

Fig. 11.21 (**a**) Algorithm proposed for HT detection in [42], (**b**) chip signatures recording the first failing frequency for pattern (Px) and capture FF (FFx), and (**c**) MDS/convex hull results for 2 HT

be tested using a power-based HT method. They argue that long paths experience less switching activity because more conditions need to be met in order to sensitize them. Therefore, power-based methods are less effective for detecting HT on these paths.

The failing frequencies for long paths are recorded in a table as shown in Fig. 11.21b. Chips are listed on rows, while the columns identify the pattern, Px, and Capture FF, FFx, of the tested paths. A multidimensional scaling (MDS) statistical method is proposed for distinguishing between delay variations introduced by process variation effects and those introduced by HT. MDS leverages PCA to map from a higher-dimensional space to a smaller space. Unlike the technique proposed in [24], however, they configure MDS to preserve signature components that represent dissimilarities introduced by HT delay anomalies in the lower-dimensional space. A 3-D convex hull is constructed using signatures from HT-free chips and outlier data points from the untrusted chips are classified as HT candidates. Their detection technique therefore is based on the GoldenSim-based or GoldenChipbased model described in reference to Fig. 11.9.

An ISCAS-89 benchmark circuit s38417 is used in their simulation experiments to validate the technique. Simulation models representing process variations are constructed by varying threshold voltage, oxide thickness, and channel length over 5% of nominal both globally and locally to model within-die variations. A total of six HT are introduced in a layout representation of the benchmark circuit with varying trigger and payload configurations. The clock frequency range and step size used for clock sweeping is set to 700 MHz to 1.5 GHz and 10 ps, respectively. The results of applying MDS and constructing a convex hull are shown in Fig. 11.21c. The detection rate for HT #1 is 64%, while the rate for HT #2 (and the remaining four HT not shown) is 100%. A similar set of results are obtained in hardware experiments using a set of 44 FPGAs.

11.5.8   A Golden Chip-Free Method for HT Detection

The authors of [36] propose the use of process control monitors (PCMs) that are designed to eliminate the need for a set of HT-free golden chips. PCMs are in-line test structures traditionally inserted by process engineers for tracking wafer-level variations in transistor parameters such as threshold voltage. The authors use the delay of a special path as a surrogate for a PCM as a silicon calibration method. Path delays from this PCM are measured from the test chips and used to improve the accuracy of the classification boundary first obtained from simulation data. This detection strategy is therefore PCM-based as discussed in Sect. 11.4.2.

The authors employ nonlinear regression and kernel mean matching techniques to learn the relationship between PCM data and side-channel fingerprints, in this case, output power measurements from a set of 40 wireless cryptographic chips which instantiate AES with and without HT. A series of "learned" boundaries are incrementally tuned as each of five statistical transformations are applied using simulation data obtained from the PCM and AES process models and from the PCM and power measurements from the test chips; Fig. 11.22a shows the first set of transformations which are derived from simulation data, illustrating transformations that take place in the shape and boundaries of the HT-free space. The remaining transformations are derived using PCM and path delay data measured from a set of HT-free chips and are illustrated in their paper. Figure 11.22b shows experimental results in which all 80 HT are correctly classified as HT infested, while only three of the HT-free chips are classified incorrectly, i.e., are false positives.

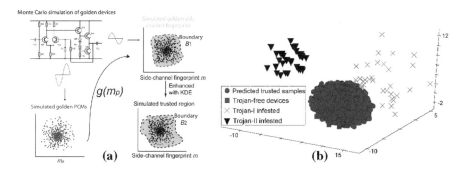

Fig. 11.22 (a) Initial simulation-based statistical transformations designed to iteratively learn the best boundaries associated with the HT-free fingerprint space for a wireless cryptographic IC from [57], (b) top three principle components from PCA analysis after application of proposed statistical learning process. All 80 HT are detected, and all but three of the HT-free chips (of 40) are classified correctly

11.5.9 HT Detection by Comparing Paths with Structural Symmetry

An HT detection method based on validating delay consistency among instances of distinct transistor-level paths with the same topology is proposed in [62]. Symmetry is defined by considering both the structural characteristics of the logic gate(s) and state assignments on its inputs under each of the vectors of an applied two-vector sequence. For example, a NAND gate exhibits symmetry in delay by having two identical pull-up paths through its two time, during a delay test. An HT detection algorithm is proposed that first identifies PMOS transistors and when input transitions are crafted to exercise each of these paths, at a times, during a delay test. An HT detection algorithm is proposed that first identifies transistor-level symmetry in the netlist or layout and then adds constraints to ATPG algorithms to test pairs of pairs that exhibit this symmetry. A self-referencing detection algorithm is proposed that compares the delays of symmetrical paths and classifies a chip as having an HT when the two path delays are not identical within a threshold.

The authors present an example of transistor-level symmetry using the ISCAS-85 c17 benchmark circuit, which is reproduced with enhancements in Fig. 11.23. The gate-level netlist of c17 is shown on the far left, while transistor-level netlists representing subsets of the netlist are shown in (a) through (d). The transistor-level diagrams are annotated with numbers to enable the NAND gates to be cross referenced to the c17 schematic. The transistor level schematics along the top and bottom rows represent the two paths that exhibit symmetry. The first two-vector sequence of the symmetry pair is labeled $\alpha 1$ and $\alpha 2$, while the second two-vector sequence is labeled $\beta 1$ and $\beta 2$. The two-vector sequences used as the tests are given as $(\alpha 1, \alpha 2)$ and $(\beta 1, \beta 2)$.

The red highlighted components show the pull-up and pull-down paths that connect the output of each NAND gate to one of the supply rails, which is determined by the logic state imposed by each of the four vectors. For example, the outputs of the three NAND gates in (a) are connected to VDD, GND, and VDD for gates labeled 2, 3, and 5. The key observation is the consistency of the highlighting between (a)–(c) and (b)–(d) and the fact that the actual gates in these pairs are different except for one of the gates. In other words, the application of the

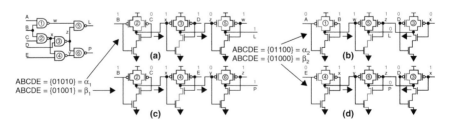

Fig. 11.23 Transistor-level symmetry illustration adapted from [56]

two vector sequences tests the same pull-up and pull-down paths in the NAND gates but do so along different paths in c17. Given that the NAND gates have identical layout structures, the path delays are expected to be nearly identical. Therefore, if an adversary inserts one or more payload gates in series with either of these paths, the delays will be different and can be flagged as a malicious modification. Simulation and FPGA results are shown to demonstrate this concept.

The delay changes introduced by global shifts in process variations (and within-die variations to some degree) are eliminated because the comparisons are made between paths on the same chip and preferably in close proximity. Therefore, none of the golden model techniques referenced in Sect. 11.4.2 for dealing with process variations is required. However, margins are needed to account for measurement noise and routing differences in the two paths; otherwise, the false-positive rates will be high. The authors indicate that finding structural symmetries in the layout and then deriving qualifying test patterns can be challenging given the large number of constraints that must be satisfied to ensure consistency in the behaviors of the pull-up and pull-down components of the tested paths. This feature can be argued as a benefit because it makes the task difficult for the adversary to carry out and then defeat the technique by inserting the HT such that the delays of symmetrical pairs of paths remain consistent.

11.5.10 HT Detection Using Pulse Propagation

A high-resolution HT detection method is proposed in [18] that is based on propagating pulses along digital logic paths. HT detection is accomplished by detecting whether pulses survive, i.e., do not die out, before reaching the capture FF where they are detected. Minimum pulse widths that allow the gates along the path to sustain the pulse are constrained by only one of the gates along the path, in particular, the gate that has the largest rise C fall time. The authors argue that this characteristic greatly enhances the HT detection sensitivity of their method to capacitive loading effects over other delay testing methods, particularly for long paths and when considering process variation effects.

Delay variations introduced by process variation effects are cumulative, and, therefore, the HT-free boundaries or margins associated with standard delay methods must be increased for longer paths, which reduce their sensitivity to small, fixed-sized variations in delay introduced by HT. On the other hand, pulses will shrink when they encounter the capacitive load of an HT and will die out at the gate that was used to determine the minimum pulse width for the path (note: this assumes the HT insertion occurs before the gate that was used to define the minimum-sized pulse). Therefore, the authors argue that HT detection sensitivity remains constant and is independent of the length of the path. The embedded components needed for pulse generation and pulse detection can be designed as shown in Fig. 11.24a, b, respectively, and these components can be shared among multiple FFs, as shown for the pipelined architecture in Fig. 11.24c.

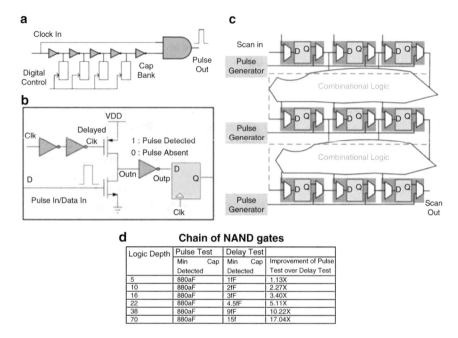

Fig. 11.24 (**a**) Proposed pulse generator, (**b**) pulse detector, (**c**) an example illustrating sharing within a pipelined architecture, and (**d**) simulation results comparing proposed technique (labeled pulse test) with standard delay test (labeled delay test) [18]

An algorithm is presented that uses n random test patterns that are each evaluated through simulation using k pulses of different widths. Test for paths that are able to propagate the pulse from a launch FF and to a capture FF is deemed valid. The authors refer to these paths as single-path sensitizable[2]. Simulations are again used to determine the minimum pulse width for each path using worst-case process models. HT are emulated on each node of every path using different capacitive loads to determine the minimum capacitance that succeeds in "killing" the propagating pulse.

The proposed method is validated using simulation experiments on a chain of NAND gates, a ripple carry adder and 4 × 4 multiplier. Process variations are modeled by changing threshold voltage (Vt) by +/− 10% globally and +/− 10% locally. The detection results for the chain of NAND gates are shown in Fig. 11.24d, with "pulse test" results corresponding to the proposed technique and "delay test" identifying results using a standard delay test strategy. The columns labeled "Min Cap Detected" represent the smallest HT capacitive load that was detectable for the paths of different lengths specified by the rows. The last column expresses the

---

[2]Single-path sensitizable refers to paths that are hazard-free robust testable, indicating all side-inputs along the path must remain constant under both applied vectors.

improvement in sensitivity of the proposed method over the standard delay test method and supports the claim that the pulse method remains sensitive to small HT even when the path length becomes very large. Similar results are obtained for the other functional units as reported in [18].

11.5.11   Chip-Centric Calibration Techniques for HT Detection

The authors of [21, 22] propose HT detection methods which use actual path delay measurements, in contrast to PCMs and other types of on-chip test structures, as a mechanism to calibrate for global shifts and within-die process variation effects[3]. These methods represent variants of the chip-centric technique described in Sect. 11.4.2. Chip-centric techniques can potentially provide higher levels of sensitivity to HT because the path delays used in the detection method also serve as the basis for calibration. Moreover, by using chip-measured path delays to shrink the HT-free space, as depicted on the right side of Fig. 11.9, such methods can also simplify the development of a simulation-based golden model, as demonstrated in [22].

The authors of [21] average path delays measured from a set of chips to reduce the adverse impact of both inter-chip and intra-die process variation effects on HT detection sensitivity. The proposed golden model is based on hardware measurements of delays from HT-free chips, i.e., design and simulation data are not used to develop the HT-free space. The data collected from the chips is multidimensional. The authors use # to symbolize the chip number, P to represent the pattern (two-vector sequence) number, N_p to represent the number of patterns, to identify functional unit outputs (Capture FFs), and Nα to represent the number of outputs. Calibration of inter-chip (global) process variation effects on path delays uses a centering operation in which the delays D under all patterns P to an output α for chip # are averaged and subtracted from each of the raw delays as given by Eq. 11.2. Therefore, the method uses the distribution of delays to each output α for calibration of the global mean shift in path delays that occurs within chip #. A second centering operation is then performed to further reduce intra-die variations which averages the globally calibrated delays to each output α across all chip outputs as given by Eq. 11.3. This chip-wide average is then subtracted from the raw path delays for a chip to provide a set of locally calibrated delays.

The ratios of two locally calibrated delays for a pattern P are used in the formulation of a golden model, each referred to as a relative performance metric, $RP_{P,\alpha,\beta}$ as given by Eq. 11.4. A matrix of relative performances is constructed for each chip #, and the mean value $\overline{RP}_{P,\alpha,\beta}$ computed across all HT-free chips, N_{GM}, is used as the references for comparison of the $RP_{P,\alpha,\beta}$ values computed from the untrusted chips. A margin referred to as the coefficient of irrelevance is proposed for dealing with false-positive HT detections. It is defined as the standard deviation

---

[3]Note, the path delay technique described in [59] is based on the same concept presented earlier in [58] which uses leakage currents.

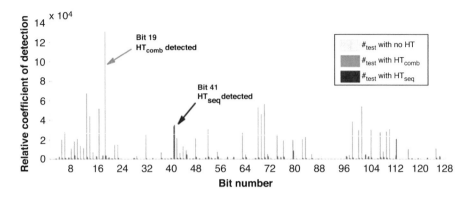

Fig. 11.25 HT detection results as presented in [21]

of the $RP_{P,\alpha,\beta}$ computed using N_{GM} HT-free chips. The threshold that bounds the HT-free space is given by Eq. 11.5 and is referred to as a distinguisher.

$$D_p(\alpha, \#) = D(P, \alpha, \#) - \frac{\sum_p (P, \alpha, \#)}{N_p} \tag{11.2}$$

$$D_{p,\alpha}(\#) = D_p(\alpha, \#) - \frac{\sum_\alpha D_p(\alpha, \#)}{N_\alpha} \tag{11.3}$$

$$RP_{P,\alpha\beta}(\#) = \frac{D_{p,\alpha}(\#)}{D_{p,\beta}(\#)} \tag{11.4}$$

$$Dg_{P,\alpha,\beta}^{\#test} = \frac{RP_{P,\alpha,\beta}(\#test) - \overline{RP}_{P,\alpha,\beta}(\#GM)}{\sigma_{P,\alpha,\beta}} \tag{11.5}$$

The technique is validated using a set of four Xilinx Spartan FPGAs programmed with an AES-128 functional unit and modified in a second design to include one combinational and one sequential HT. The golden model is built using delays measured from the AES-128 without the HT. The single-clock scheme (or clock sweeping from Sect. 11.4.1.3) is used with a step size of 35 ps and a frequency range from 100 MHz to 121.2 MHz. Test vector selection is performed randomly, i.e., no test vector generation strategy is proposed. A set of 50 patterns (plaintexts) are used as the test vector set, and paths from all 128 bits of the AES are monitored. Path delays shorter than 8.25 ns (1/121.2 MHz) are ignored. The authors report on a subset of the distinguishers, in particular, the distinguishers which produced the maximum value for each of the 128 outputs when computed using the HT-free data and data from the two HT experiments. The results are shown in Fig. 11.25, with highlights indicating the outputs that provide the highest levels of confidence in detecting the two HT.

Although the technique proposed in [21] is demonstrated to work well, the averaging techniques that the authors employ do not deal directly with intra-chip process variations. The technique presented in [22] (discussed below), on the other hand, averages delays across chips for each path and two-vector sequence, instead of across vectors and outputs. Within-die variations have been shown to have a significant random component in each chip instance [16], and, therefore, a path-by-path averaging strategy is likely to be more effective in reducing unwanted intra-chip variations. Moreover, the strategy proposed in [21] only calibrates for the global shift in the mean values of path delays introduced by inter-chip process variation effects. The technique described in the following also considers scaling effects.

A chip-averaging HT detection method that calibrates for both intra-chip and inter-chip process variations and measures path delays using an on-chip time-to-digital converter (TDC) is proposed in [22]. The TDC was described earlier in reference to Fig. 11.8 in Sect. 11.4.1.3. The TDC provides approx. 25 ps of timing resolution, is very fast, e.g., no clock strobing or clock sweeping operation is required, and can be multiplexed and shared across a large number of the functional unit outputs. The method is also classified as chip-centric but unlike [21] does not depend on a set of golden chips. Rather, a golden simulation model is used to characterize the HT-free space. The development of the golden model requires only a single nominal simulation to be run for each of the applied two-vector sequences, and therefore the approach significantly reduces the level of effort and time required over previously proposed simulation-based golden model approaches.

This is possible because the calibration processes are geared toward deriving a nominal chip-averaged-delay (CAD) value for each path from hardware data, and therefore, process variation effects do not need to be accounted for in the golden model.

Calibration and chip averaging are designed to reduce performance differences and the adverse effects of process variations on delay while preserving any type of systematic variation that shows up in all (or a large subset) of the tested chips. Chip averaging leverages a key difference between random process variations and HT anomalies; random variations average to 0, while HT anomalies introduce systematic differences that survive the averaging process.

The authors validate the method using data collected from 44 copies of an ASIC fabricated in a 90 nm technology which has two exact copies of the layout of an AES functional unit, one representing the original design and one with five embedded HT. A layout of the chip showing the two copies of the AES and four instances of the TDC is shown in Fig. 11.26a. The two 8-to-1 multiplexers shown in the block diagram of the TDC from Fig. 11.8 connect to 15 of the 128 outputs of the AES (the 16th input is connected to the Clk). The two copies of the TDC in each AES instance allow signals propagating to 30 of the outputs of AES to be timed against the Clk.

The calibration process used to reduce inter-chip process variations is carried out in advance of the HT detection procedure on each chip separately. Similar to the HT detection process, calibration involves measuring delays from paths of various lengths within the functional unit on each chip. Unlike HT detection, the

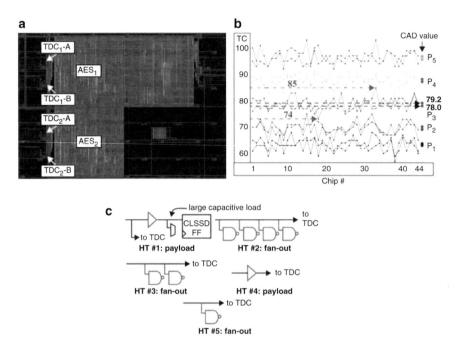

Fig. 11.26 (**a**) Chip layout showing two copies of the AES [10], (**b**) TCs from 44 chips with CAD values shown on far right, and (**c**) configurations of HT added to AES2

goal of calibration is to tune the control signals, Cal0 and Cal1, of the TDC as a means of shifting (and scaling) the delay distribution obtained for each chip to a fixed mean value. From Sect. 11.4.1.3, the output of the TDC is a thermometer code (TC), i.e., an integer value between 0 and 120, that represents the relative delay difference between the Clk and the path being tested. The fixed mean value is set to the halfway point (60). By using the same fixed mean value for all chips, this process effectively standardizes the TCs, thereby eliminating most of the delay variations introduced by chip-to-chip process variation effects. The chip-averaging technique is designed to remove the remaining intra-chip variation that exists in the path delays. Once the TDC is calibrated, a set of TDF-based vectors designed to test each possible HT site in a hazard-free fashion is applied to the chips. The HT detection method is applied once data from all or a large sample, e.g., 50 or more, chips is collected. A chip-averaged-delay (CAD) value is computed for each tested path by averaging the TC delays obtained from all chips. The CAD averages and ideally eliminates random within-die variations, making it possible to observe very small systematic differences which occur in the chip values but are not present in a Spice-level simulation of the nominal model. As an illustration, Fig. 11.26b plots the raw TC values for 5 HT-free paths of different lengths. The x-axis lists the chip, 1–44, for each TC value on the y-axis. The two curves represent the data collected from each of the two nearly identical AES instantiations shown in Fig. 11.26a. The variations in the data points across chips and between the AES instantiations are what remain after calibration and are attributed to intra-die variations and measurement noise. The CAD values for each of the five paths are shown as the last point of the waveforms on the far right. The chip-averaging effect is reflected in the "closeness" of the two points computed from the 44 chips of each AES instantiation. The layout of the two AES instantiations is identical, and, therefore, ideally, the CAD values should be superimposed. Although this is not the case, the CAD values are closer than most of raw TC values for any given chip. This reduces the boundaries associated with the HT-free space, which in turn, improves the HT detection sensitivity of the proposed method. The authors validate the detection sensitivity of the method by measuring the delay anomalies introduced by five layout-inserted HT. Figure 11.26c gives schematic-level diagrams illustrating the structure and insertion points of the HT, which are highlighted in red. Four fanout HT and one series-inserted HT are added to the layout of AES2 by replacing filler cells and connecting the inputs and outputs of the HT as shown by the schematic. A nominal simulation model of the AES layout and TDC is created using Mentor Graphics Calibre XRC extractor and the foundry-provided models for the 90 nm technology in which the chips were fabricated. Transient simulations using Cadence Spectre are carried out to obtain the TC values associated with the nominal model.

The graphs shown in Fig. 11.27 plot two sets of results, (a) plots the simulation nominal model data against the HT-infested AES2 data, while (b) plots the simulation data against the HT-free AES1 data for the same 20 paths. The y-axis plots a DCAD value, which is simply the difference between the simulation TC value and hardware-derived CAD values. HT that introduce larger anomalies therefore generate larger DCAD values. The paths are sorted left to right according to the magnitude of the HT delay anomaly, with the largest DCAD values on the left. The red curves represent data collected from paths that include one of the HT shown in Fig. 11.26c, while the black curves represent data from HT-free paths. The displacement of the red curves upward with respect to the black curves in (a)

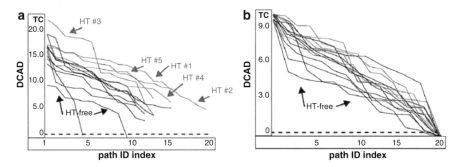

Fig. 11.27 (**a**) DCAD values of golden simulation model against HT-infested AES2 and (**b**) DCAD value of golden simulation model against HT-free AES1 from [36]

portrays the presence of the delay anomaly introduced by the HT. The curves in (b), on the other hand, show that the DCAD values for these same paths from AES1 (which does not include the HT) are interleaved with the HT-free (black) curves.

11.6   Power Consumption Based Side-Channel Analysis for HT Detection Approaches

Side-channel signals are typically analog in nature and can provide detailed, high-resolution information about the internal timing and regional signal behavior of the IC. For example, IDDT measurements reflect performance characteristics of individual gates as logic signals propagate along one or more paths in the circuit. This type of temporal information can be reverse engineered and compared with simulation-generated data to validate the structural characteristics of the fabricated layout, i.e., to ensure the chip is consistent with the golden model representation described by the design data.

As discussed earlier in Sect. 11.3.1.2, other common methods of power analysis based Trojan detection include current analysis during static and transient states. The static current analysis technique can detect the inactivated Trojans that are large enough to impact the change in the current drawn because of addition or modification of reconvergent paths and gates. On the other hand, the transient current analysis allows the activated Trojan that changes the current during operation [8, 9] and [1, 2, 10, 25, 26, 28, 30, 43].

The authors of [41] leverage correlations between maximum operating frequency, Fmax, and transient current, IDDT, as a mechanism to enhance the HT detection sensitivity of IDDT. Multiparameter side-channel analysis refers to the joint analysis of two or more circuit parameters, such as power and delay, as a means of accounting for process variation effects or to provide higher levels of confidence that an HT exists through corroborative evidence, or a lack thereof, from multiple signal sources. This concept is portrayed in Fig. 11.28a which plots IDDT against Fmax. Here, simulation experiments are used to show an embedded HT effect

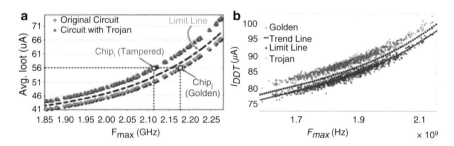

Fig. 11.28 (a) Correlations between IDDT and Fmax distinguish HT-free and HT-infested chips in the presence of process noise, (b) similar analysis with random intra-die variations added [41]

IDDT because of additional HT switching activity but does not impact Fmax. The mismatch in the correlation of IDDT and Fmax allows the HT to be identified in the "tampered" IDDT curve that would otherwise not be possible. Fmax is effectively used to track process variation effects. The distinction is blurred to some degree with the addition of random within-die variations as shown in Fig. 11.28b, but the correlation and benefit provided by Fmax remain apparent in the displacement and separation of the red (HT) and blue (HT-free) data points. The authors note that any path or set of paths can be used for the correlation analysis to make it nearly impossible for the adversary to defeat the technique.

The authors propose a test vector generation strategy that first partitions the multi-module design into nonoverlapping functional blocks as a mechanism to amplify the HT IDDT contribution (signal) over normal background IDDT (noise). Vectors optimized to target HT nodes are selected and directed at testing one of the blocks while simultaneously minimizing activity in other functional blocks. The test vectors for IDDT and a separate set for Fmax are used in the proposed test flow to optimize correlations as shown in Fig. 11.28. Simulation and FPGA results are presented which validate their approach.

Wilcox [58] proposes a MSP technique, in combination with a power signal calibration and a chip averaging method to reduce the hardware Trojan signal-to-noise ratio to capture small leakage anomalies introduced by Trojans and separate it from large leakage current anomalies caused by the defects such as broad area leakage current and shortening defects. The technique measures steady state leakage current IDDQ from multiple (16 VDD and GND ports) topology distributed power ports on the chip and applies chip averaging method (as discussed in Sect. 11.5.10) for eliminating within-die variation and improving hardware Trojan signal-to-noise detection sensitivity of the statistical based detection method. The layout of the chip AES and FPU engines is shown in Fig. 11.22. The instrumentation setup is shown in Fig. 11.29. Large low resistance M9 wires are routed from a set of VDD and GND pads, to a 4×4 grid of tap points distributed at $250\,\mu$ m intervals across the macros [17].

Fig. 11.29 Instrumentation setup

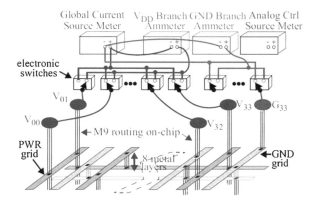

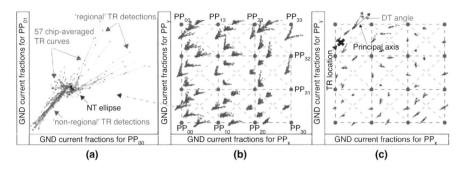

Fig. 11.30 Chip averaged GND, leak pattern 1 data. (**a**) PP pairing no Trojan eclipse and all Trojan data (**b**) All PP pairing showing ellipse and all 57 Trojan data curves (**c**) All PP pairings showing ellipse and Trojan data #1 curves [17]

The currents are measured from the adjacent pairings of power ports and are used in combination to outlier analysis to distinguish random defects and the leakage anomalies introduced by the hardware Trojans. The technique uses golden model that can be from the simulation models or from the Trojan free chips. The ellipse statistical limits are derived from the golden reference chips data using principle component analysis. The chip averaging technique is applied [21, 58], which is a very effective tool for identifying systematic current anomalies that occur across chips. PCA results after the chip averaging shows that the Trojan-free data is enclosed in the ellipse whereas the Trojan data is seen as outliers as shown in Fig. 11.30. Histogram results validate their approach, with zero false positive detection for all the leakage patterns.

11.7   Summary

Hardware Trojans (HT) represent a serious threat and a significant challenge. Side-channel techniques, such as power and delay analysis, can be argued as the most sensitive and cost-effective strategies for detecting HT. This chapter surveyed a wide variety of delay-based approaches that have been proposed over the last decade. Important technical aspects and distinctions that characterize the proposed HT detection methods can be summarized as follows:

- Path delay measurement strategy for obtaining precise measurements of path delays:
    Clock sweeping implemented by adjusting the frequency of applied clock
    Two clock approaches which tune the phase between launch and capture clocks (clock strobing)
    On-chip, embedded test structures which create (tunable) delay chains

- Side-channel signals are analog in nature and can provide detailed, high-resolution information about the internal timing and regional signal behavior of the IC
  IDDT and IDDQ techniques are used for transient and steady state currents
- Test stimulus strategies for HT detection: Random vectors
  Vectors generated using the traditional transition fault delay (TDF) model
  Vectors generated from a pseudo-TDF model targeting shortest sensitizable paths
  Pulse stimulus-based techniques
- Approaches to account for process variation effects, both chip-to-chip and within-die: HT-free space created from process simulation models
  HT-free space created from data collected from golden (HT-free) chips (which are validated using destructive delayering techniques)
  Simulation-derived HT-free space calibrated with hardware data from process control monitors (PCMs), ring oscillators (ROs), critical paths, etc.
  Techniques which average path delays measured from (untrusted) chips and compared against (nominal) simulation models or golden HT-free chips
  Techniques which correlate multiple side-channel signals
- Design-for-trust additions, modifications, and analyses to support HT detection methods: Techniques which create ROs from functional unit paths
  Techniques which add a distributed set of ROs designed to detect HT switching activity
  Methods designed to find structural symmetry in path delays for comparison
  Techniques which add symmetrical components to enable calibration using chip data
- Statistical HT detection methods: Simple thresholding and linear regression-based methods
  Advanced statistical analysis techniques which employ nonlinear regression, kernel mean matching, principle component analysis, multidimensional scaling, and convex hull construction
  Ad hoc statistical techniques which leverage path delay differences, ratios, and other mathematical transformations

Taken collectivity, three critical features emerge as requirements for a fully specified and effective HT detection method:

- First, traditional manufacturing test methods are not capable of providing precise measurements of path delays, which is a requirement of nearly all proposed HT detection methods. Therefore, a paradigm shift is required in the way path delaytesting is carried out by automatic test equipment and/or in the capabilities of design-for-testability support structures included on the chip. Several low-cost embedded test structures were described that support high-resolution on-chip measurements of path delays.
- Second, both within-die and chip-to-chip process variations pose significant limits on HT detection sensitivities and must be dealt with in a cost-effective manner. Golden model-based methods must be based on realistic assumptions regarding the availability of golden chips and the amount of simulation time and

effort required to define the boundaries of a multidimensional HT-free space. Golden model-free methods must have validation techniques to guard against subversion by the adversary.

- Third, a low-cost test vector generation strategy must be developed that is effective at detecting subtle HT loading effects and which also provides high levels of HT coverage while minimizing test cost.

Achieving all these goals is very challenging, but the commercial acceptance of path delay testing as a mainstream HT detection strategy critically depends on low-cost solutions to all three of these technical domains.

References

1. J. Aarestad, D. Acharyya, R. Rad, J. Plusquellic, Detecting Trojans though leakage current analysis using multiple supply pad IDDQs. Trans. Inf. Forensics Secur. **5**(4), 893–904 (2010)
2. D. Agrawal, S. Baktir, D. Karakoyunlu, P. Rohatgi, B. Sunar, Trojan detection using IC fingerprinting, in *Symposium on Security and Privacy* (2007), pp. 296–310
3. M. Banga, M. Hsiao, A region based approach for the detection of hardware Trojans, in *Workshop on Hardware-Oriented Security and Trust* (2008), pp. 40–47
4. M. Banga, M. Hsiao, A novel sustained vector technique for the detection of hardware Trojans, in *International Conference on VLSI Design* (2009), pp. 327–332
5. M. Banga, M. Chandrasekar, L. Fang, M. Hsiao, Guided test generation for isolation and detection of embedded Trojans in ICs, in *Great Lakes Symposium on VLSI* (2008), pp. 363–366
6. M. Beaumont, B. Hopkins, T. Newby, Hardware Trojans - prevention, detection, countermeasures (Department of Defense, Australian Government, Canberra, 2011)
7. S. Bhunia, M. Tehranipoor (eds.), *Hardware Trojan War: Attacks, Myths, and Defenses* (Springer, Berlin, 2018)
8. S. Bhunia, M. Abramovici, D. Agrawal, P. Bradley, M.S. Hsiao, J. Plusquellic, M. Tehranipoor, Protection against hardware Trojan attacks: towards a comprehensive solution. Des. Test 30(3), 6–17 (2013)
9. S. Bhunia, M. Hsiao, M. Banga, S. Narasimhan, Hardware Trojan attacks: threat analysis and countermeasures. Proc. IEEE **102**(8), 1229–1247 (2014)
10. M. Bushnell, V.D. Agrawal, Essentials of electronic testing for digital, memory, and mixed signal VLSI circuits, vol. 17 (Springer, Basel, 2000)
11. J.L. Carter, V.S. Iyengar, B.K. Rosen, Efficient test coverage determination for delay faults, in *International Test Conference* (1987), pp. 418–427
12. B. Cha, S.K. Gupta, Efficient Trojan detection via calibration of process variations, in *Asian Test Symposium* (2012)
13. B. Cha, S.K. Gupta, Trojan detection via delay measurements: a new approach to select paths and vectors to maximize effectiveness and minimize cost, in *Design, Automation & Test in Europe* (2013)
14. R.S. Chakraborty, S. Narasimhan, S. Bhunia, Hardware Trojan: threats and emerging solutions, in *International High Level Design Validation and Test Workshop* (2009), pp. 166–171
15. R.S. Chakraborty, F. Wolff, S. Paul, C. Papachristou, S. Bhunia, MERO: a statistical approach for hardware Trojan detection, in *Workshop on Crytographic Hardware and Embedded Systems* (2009), pp. 396–410
16. W. Che, M. Martin, G. Pocklassery, V.K. Kajuluri, F. Saqib, J. Plusquellic, A privacy preserving, mutual PUF-based authentication protocol. Cryptography **1**(1) (2016)

17. M. Chen, P. Mishra, Property learning techniques for efficient generation of directed tests. IEEE Trans. Comput. **60**(6), 852–864 (2011)

18. S. Deyati, B.J. Muldrey, A. Singh, A. Chatterjee, High resolution pulse propagation driven trojan detection in digital logic: optimization algorithms and infrastructure, in *Asian Test Symposium* (2014), pp. 200–205

19. D. Du, S. Narasimhan, R.S. Chakroborty, S. Bhunia, Self-referencing: a scalable side-channel approach for hardware Trojan detection, in *Cryptographic Hardware and Embedded Systems* (2010), pp. 173–187

20. D. Ernst, S. Das, S. Lee, D. Blaauw, T. Austin, T. Mudge, N.S. Kim, K. Flautneret, Razor:circuit-level correction of timing errors for low-power operation. Micro **24**(6), 10–20 (2004)

21. I. Exurville, L. Zussa, J.-B. Rigaud, B. Robisson, Resilient hardware Trojans detection based on path delay measurements, in *International Symposium on Hardware-Oriented Security and Trust* (2015), pp. 151–156

22. D. Ismari, C. Lamech, S. Bhunia, F. Saqib, J. Plusquellic, On detecting delay anomalies introduced by hardware Trojans, in *International Conference on Computer-Aided Design* (2016)

23. N. Jacob, D. Merli, J. Heyszl, G. Sigl, Hardware Trojans: current challenges and approaches. IET Comput. Digit. Tech. **8**(6), 264–273 (2014)

24. Y. Jin, Y. Makris, Hardware Trojan detection using path delay fingerprint, in *Workshop on Hardware-Oriented Security and Trust* (2008), pp. 51–57

25. J. Kalisz, Review of methods for time interval measurements with picosecond resolution. Metrologia **41**(1), 17–32 (2003)

26. D. Karaklajic, J.-M. Schmidt, I. Verbauwhede, Hardware designer's guide to fault attacks. Trans. VLSI Syst. **21**(12), 2295–2306 (2013)

27. R. Karri, J. Rajendran, K. Rosenfeld, M. Tehranipoor, Trustworthy hardware: identifying and classifying hardware Trojans. Computer **43**(10), 39–46 (2010)

28. P. Kocher, J. Jaffe, B. Jun, Differential power analysis, in *Advances in Cryptology* (Springer, Berlin, 1999)

29. C. Lamech, J. Plusquellic, Trojan detection based on delay variations measured using a high precision, low-overhead embedded test structure, in *Hardware-Oriented Security and Trust* (2012), pp. 75–82

30. C. Lamech, J. Aarestad, J. Plusquellic, R.M. Rad, K. Agarwal, REBEL and TDC: embedded test structures for regional delay measurements, in *International Conference on Computer-Aided Design* (2011), pp. 170–177

31. J. Li, J. Lach, Negative-skewed shadow registers for at-speed delay variation characterization, in *International Conference on Computer Design* (2007), pp. 354–359

32. J. Li, J. Lach, At-speed delay characterization for ic authentication and Trojan horse detection, in *Workshop on Hardware-Oriented Security and Trust* (2008), pp. 8–14

33. M. Li, A. Davoodi, M. Tehranipoor, A sensor-assisted self-authentication framework for hardware Trojan detection, in *Design, Automation & Test in Europe Conference* (2012)

34. C.J. Lin, S.M. Reddy, On delay fault testing in logic circuits. Trans. Comput.-Aid Des. CAD **6**(5), 694–703 (1987)

35. J.-J. Liou, K.-T. Cheng, D.A. Mukherjee, Path selection for delay testing of deep sub-micron de-vices using statistical performance sensitivity analysis, in *VLSI Test Symposium* (2000)

36. Y. Liu, K. Huang, Y. Makris, Hardware Trojan detection through golden chip-free statistical side-channel fingerprinting, in *Design Automation Conference* (2014), pp. 1–6

37. E. Love, Y. Jin, Y. Makris, Proof-carrying hardware intellectual property: a pathway to trust-ed module acquisition. Trans. Inf. Forensics Secur. **7**(1), 25–40 (2012)

38. A.K. Majhi, V.D. Agrawal, Delay fault models and coverage, in *International Conference on VLSI Design* (1998)

39. Y.K. Malaiya, R. Narayanaswamy, Modeling and testing for timing faults in synchronous sequential circuits. Des. Test Comput. **1**(4), 62–74 (1984)

40. S.R. Nassif, Design for variability in DSM technologies, in *International Symposium on Quality Electronic Design* (2000)
41. S. Narasimhan, D. Du, R.S. Chakraborty, S. Paul, F. Wolff, C. Papachristou, K. Roy, S. Bhunia, Multiple-parameter side-channel analysis: a non-invasive hardware trojan detection approach, in *International Symposium on Hardware-Oriented Security and Trust* (IEEE, Anaheim, 2010), pp. 13–18
42. M. Potkonjak, A. Nahapetian, M. Nelson, T. Massey, Hardware Trojan horse detection using gate-level characterization, in *Design Automation Conference* (2009), pp. 688–693
43. R. Rad, J. Plusquellic, M. Tehranipoor, Sensitivity analysis to hardware Trojans using power supply transient signals, in *Workshop on Hardware-Oriented Security and Trust* (2008), pp. 3–7
44. D. Rai, J. Lach, Performance of delay-based Trojan detection techniques under parameter variations, in *International Workshop Hardware-Oriented Security and Trust*, 2009, pp. 58–65
45. J. Rajendran, V. Jyothi, O. Sinanoglu, R. Karri, Design and analysis of ring oscillator-based de-sign-for-trust technique, in *VLSI Test Symposium* (2011), pp. 105–110
46. J. Rajendran, O. Sinanoglu, R. Karri, Is split manufacturing secure?, in *Proceedings of IEEE Design, Automation and Test in Europe Conference & Exhibition (DATE)*, 2013, Grenoble, France, 18–22 March (2013), pp. 1259–1264
47. H. Salmani, M. Tehranipoor, Vulnerabilities analysis of a circuit layout to hardware trojan insertion, in *IEEE Transactions on Information Forensics and Security* (2016)
48. H. Salmani, M. Tehranipoor, J. Plusquellic, A layout-aware approach for improving localized switching to detect hardware Trojans in integrated circuits, in *International Workshop on Information Forensics and Security* (2010)
49. H. Salmani, M. Tehranipoor, J. Plusquellic, A novel technique for improving hardware Trojan detection and reducing Trojan activation time. Trans. VLSI Syst. **20**(1), 112–125 (2012)
50. J. Soden, R. Anderson, C. Henderson, Failure analysis tools and techniques - magic, mystery, and science, in *International Test Conference, Lecture Series II "Practical Aspects of IC Diagnosis and Failure Analysis: A Walk through the Process"* (1996), pp. 1–11
51. G.L. Smith, Model for delay faults based upon paths, in *International Test Conference* (1985), pp. 342–349
52. M. Tehranipoor, F. Koushanfar, A survey of hardware Trojan taxonomy and detection. IEEE Des. Test Comput. **27**(1), 10–25 (2010)
53. M. Tehranipoor, C. Wang (eds.), *Introduction to Hardware Security and Trust*. Springer, New York (2011)
54. X. Wang, M. Tehranipoor, J. Plusquellic, Detecting malicious inclusions in secure hardware: challenges and solutions, in *International Workshop on Hardware-Oriented Security and Trust* (2008), pp. 15–19
55. X. Wang, M. Tehranipoor, R. Datta, Path-RO: a novel on-chip critical path delay measurement under process variations, in *International Conference on Computer-Aided Design* (2008)
56. S. Wei, M. Potkonjak, Malicious circuitry detection using fast timing characterization via test points, in *Symposium on Hardware-Oriented Security and Trust* (2013)
57. S. Wei, K. Li, F. Koushanfar, M. Potkonjak, Provably complete hardware Trojan detection using test point insertion, in *International Conference on Computer-Aided Design* (2012), pp. 569–576
58. I. Wilcox, F. Saqib, J. Plusquellic, GDS-II Trojan detection using multiple supply pad VDD and GND IDDQs in ASIC functional units, in *International Symposium on Hardware-Oriented Security and Trust* (2015)
59. F. Wolff, C. Papachristou, S. Bhunia, R.S. Chakraborty, Towards Trojan-free trusted ICs: problem analysis and detection scheme, in *Design, Automation and Test in Europe* (2008)
60. K. Xiao, X. Zhang, M. Tehranipoor, A clock sweeping technique for detecting hardware Trojans impacting circuits delay. Des. Test **30**(2), 26–34 (2013)

61. K. Xiao, D. Forte, Y. Jin, R. Karri, S. Bhunia, M. Tehranipoor, Hardware Trojans: lessons learned after one decade of research. ACM Trans. Des. Autom. Electron. Syst. **22**(1), 6:1–6:23 (2016)
62. N. Yoshimizu, Hardware Trojan detection by symmetry breaking in path delays, in *International Symposium on Hardware-Oriented Security and Trust* (2014), pp. 107–111
63. X. Zhang, M. Tehranipoor, RON: an on-chip ring oscillator network for hardware Trojan detection, in *Design and Test in Europe* (2011)

Chapter 12
The Future of Security Validation and Verification

12.1 Introduction

System-on-Chip (SoC) is the brain behind the computing devices today. Unlike microcontroller based designs in the past, even resource constrained Internet-of-Things (IoT) devices nowadays incorporate one or more complex SoCs. A typical SoC consists of multiple intellectual property (IP) cores including processor, memory, network-on-chip, controllers, converters, input/output devices, etc. Drastic increase in SoC complexity has led to significant increase in SoC design and validation complexity. Reusable hardware IP based SoC design has emerged as a pervasive design practice in the industry to dramatically reduce design and verification cost while meeting aggressive time-to-market constraints. Growing reliance on these pre-verified hardware IPs, often gathered from untrusted third-party vendors, severely affects the security and trustworthiness of SoC computing platforms. Hardware-level vulnerabilities should be fixed before deployment since it affects the overall system security. Based on Common Vulnerability Exposure (CVE-MITRE) estimates, if hardware-level vulnerabilities are removed, the overall system vulnerability will be reduced by 43% [12, 16]. Given the widespread acceptance of SoC designs in the electronic industry, it is critical to ensure their correctness from both functional and security perspectives.

12.2 Summary

This book provided a comprehensive overview of state-of-the-art security validation and verification techniques for designing secure and trustworthy SoCs [2, 13–15, 18, 21–24, 38]. The previous chapters covered SoC security verification using a wide variety of techniques including formal verification, simulation-based validation,

© Springer Nature Switzerland AG 2020   273
F. Farahmandi et al., *System-on-Chip Security*,
https://doi.org/10.1007/978-3-030-30596-3_12

side-channel analysis, and machine learning. The topics covered in this book can be broadly divided into the following four categories.

12.2.1   Introduction to Security Validation

The first two chapters introduced SoC security vulnerabilities due to potentially untrusted semiconductor supply chain and presented the fundamental challenges in detecting and mitigating SoC security vulnerabilities. The third chapter described security metrics and benchmarks.

- Chapter 1 introduced the role of semiconductor supply chain in today's SoC design methodology. Specifically, it highlighted various types of potential threats during different design stages. It provided an overview of various SoC security vulnerabilities including malicious implants (e.g., hardware Trojans).
- Chapter 2 outlined the fundamental challenges in verifying SoC security vulnerabilities. Specifically, it highlighted the limitations of applying the existing functional validation methods for security verification. For example, in case of malicious implants, it highlighted the difficulty of detecting Trojans due to rareness of its trigger conditions. Similarly for finite state machines, it highlighted the exponential search space of finding illegal states and transitions. These challenges form the stepping stone for the security validation approaches described in the subsequent chapters.
- Chapter 3 described security metrics and benchmarks, which are vital components in evaluating the trustworthiness of SoCs. It described both static and dynamic benchmarks and highlighted the importance of dynamic benchmarks to compare the quality of security verification techniques. It presented a Trojan insertion tool that can generate dynamic benchmarks by inserting a wide variety of Trojans.

12.2.2   Formal Verification of Security Vulnerabilities

The next four chapters described various formal verification techniques to detect SoC security vulnerabilities.

- Chapter 4 presented an automated methodology for anomaly detection in complex arithmetic circuits. It used the remainder produced by equivalence checking methods to generate directed tests that are guaranteed to activate the source of the malicious functionality in the design. It used the generated tests to localize the source of the anomaly and find suspicious areas in the design [17, 19].
- Chapter 5 described an automated approach to localize hardware Trojans in third-party IPs using symbolic algebra. The technique is based on extracting polynomials from gate-level implementation of the untrustworthy component

and comparing them with specification polynomials. It also presented a greedy test generation method to activate hardware Trojans [21, 22].

- Chapter 6 highlighted the importance of securing finite state machines (FSMs) since any unexpected functionality can endanger the integrity of the SoC design. This chapter presented an approach to formally detect anomalies in FSMs using symbolic algebra. It models both FSM specification and implementation using polynomials and checks the equivalence using Gröbner basis theory [18].
- Chapter 7 deals with SoC security verification using property checking. Defining a comprehensive set of security properties allows detection of security violations at early stages of design cycle. It discussed efficient property checking techniques for SoC security verification.

12.2.3 Security Validation Using Simulation and Learning Methods

The next two chapters deal with SoC security validation using simulation-based validation as well as machine learning.

- Chapter 8 focused on efficient test generation approaches for hardware Trojan detection. Test generation for Trojan detection is challenging since Trojans are stealthy in nature and it is hard to activate the triggering conditions. This chapter presented several test generation approaches including ATPG, statistical test generation, and directed test generation using formal methods [7, 8, 10, 11, 29, 32, 35, 40]. It also presented a scalable test generation technique using an effective combination of concrete simulation and symbolic execution [1, 3, 31, 33, 41].
- Chapter 9 discussed machine learning techniques for detection of hardware Trojans. When suitable features are extracted and the model is sufficiently trained, machine learning algorithms can find patterns to differentiate between Trojan-infected and Trojan-free designs that are beyond the capability of human analysis or side-channel signatures.

12.2.4 Security Validation Using Side-Channel Analysis

The next two chapters presented SoC security validation utilizing side-channel signatures such as dynamic current and path delay analysis [27, 28, 30, 31].

- Chapter 10 presented a framework for scalable test generation, which can significantly improve the Trojan detection sensitivity in side-channel analysis based Trojan detection. The approach aims at statistically increasing switching activity in the suspicious regions while reducing the switching activity in the rest of the design. As a result, it can amplify the Trojan effect in the presence of large process variations, making it suitable for Trojan detection using dynamic current analysis.

- Chapter 11 surveyed a wide variety of delay-based side-channel analysis approaches for detection of hardware Trojans. It described the requirements of path-delay-based Trojan detection methods. It summarized a wide range of proposed approaches and evaluated their strengths and weaknesses.

12.3   Future Directions

In spite of extensive research efforts in developing scalable and automated security validation techniques over the years, there are still many challenges remain to design secure and trustworthy SoCs. A promising direction in security validation is to effectively utilize machine learning. While directed tests can check for known vulnerabilities, machine learning can extend the scope for both known and unknown (e.g., known vulnerabilities with minor or major variations) SoC security vulnerabilities. Existing data in verification environment traces can be clustered into several buckets such that each bucket contains traces that have failed as the result of the same cause. Therefore, wasted effort in debugging of known security failures can be avoided. The future research will explore machine learning techniques for different classes of SoC vulnerabilities and develop suitable feature extraction and training mechanisms. Since an adversary will be aware of the machine learning models for security verification, it is likely that an adversary will try to tamper the machine learning models (in addition to tampering SoC IPs). In such an adversarial machine learning scenario, there will be significant research effort in developing techniques to verify robustness of machine learning models against adversarial attacks.

While design-time security validation techniques can detect certain types of vulnerabilities, it is infeasible to remove all possible vulnerabilities during pre-silicon security validation. Due to observability constraints in fabricated SoCs, post-silicon security validation approaches have to rely on observability-aware techniques (e.g., observability-aware test generation [5, 20]). In order to defend against attacks on debug infrastructure (e.g., trace-buffer attack [25, 26]), there will be renewed effort in developing secure debug architecture design (e.g., security-aware signal selection [4, 42, 43]). The future research needs to employ both post-silicon security validation [37] and runtime security monitoring techniques. For example, system-level security monitoring approaches can utilize dedicated hardware design to check the operation of an embedded processor instruction-by-instruction. Any deviations from the expected behavior (which may come from runtime attacks) can be treated as a security threat [6, 39, 44]. Future approaches need to make such security mechanisms smart and energy efficient. There will also be a significant emphasis on detecting unintentional vulnerabilities. For example, existing electronic design automation (EDA) tools can introduce several types of security vulnerabilities. There will be big push in designing EDA tools that are guaranteed to produce correct designs as per specification—nothing more, nothing less. Clearly, someone has to deal with the hard task of developing a complete and accurate specification at the first place, and keep updating the specification when there are any changes [9, 34, 36]. Finally, security validation tools need to check for various security vulnerabilities across different phases in the design cycle.

There are several classes of hardware security vulnerabilities such as access privileges, buffer errors, resource management, information leakage, numeric errors, crypto errors, and code injection. These vulnerabilities coupled with a wide variety of software and firmware attacks threaten the security and integrity of computing systems. A vast majority of hardware security research have focused on malicious modifications such as hardware Trojans. The future research needs to address diverse hardware security vulnerabilities and develop effective countermeasures. This book highlighted security verification methods using formal verification, simulation-based validation, machine learning as well as side-channel analysis. Each of these methods has its inherent merits and demerits. There will be a need for developing hybrid approaches combining the inherent advantages of different security verification methods to detect a wide variety of security vulnerabilities in emerging SoCs.

References

1. A. Ahmed, P. Mishra, QUEBS: qualifying event based search in concolic testing for validation of RTL models, in *IEEE International Conference on Computer Design (ICCD)* (2017), pp. 185–192
2. A. Ahmed, F. Farahmandi, Y. Iskander, P. Mishra, Scalable hardware Trojan activation by interleaving concrete simulation and symbolic execution, in *IEEE International Test Conference (ITC)* (2018)
3. A. Ahmed, F. Farahmandi, P. Mishra, Directed test generation using concolic testing of RTL models, in *Design Automation and Test in Europe (DATE)* (2018), pp. 1538–1543
4. K. Basu, P. Mishra, Efficient trace signal selection for post silicon validation and debug, in *International Conference on VLSI Design* (2011), pp. 352–357
5. K. Basu, P. Mishra, P. Patra, Observability-aware directed test generation for soft errors and crosstalk faults, in *International Conference on VLSI Design* (2013), pp. 291–296
6. S. Charles, Y. Lyu, P. Mishra, Real-time detection and localization of DoS attacks in NoC based SoCs, in *Design Automation and Test in Europe (DATE)* (2019)
7. M. Chen, P. Mishra, Functional test generation using efficient property clustering and learning techniques. IEEE Trans. Comput. Aided Des. Integr. Circuits Syst. **29**(3), 396–404 (2010)
8. M. Chen, P. Mishra, Property learning techniques for efficient generation of directed tests. IEEE Trans. Comput. **60**(6), 852–864 (2011)
9. M. Chen, P. Mishra, Assertion-based functional consistency checking between TLM and RTL models, in *International Conference on VLSI Design* (2013), pp. 320–325
10. M. Chen, P. Mishra, D. Kalita, Automatic RTL test generation from SystemC TLM specifications. ACM Trans. Embed. Comput. Syst. **11**(2), article 38 (2012)
11. M. Chen, X. Qin, P. Mishra, Learning-oriented property decomposition for automated generation of directed tests. J. Electr. Test. **30**(3), 287–306 (2014)
12. Common weakness enumeration (2017). https://cwe.mitre.org/

13. J. Cruz, Y. Huang, P. Mishra, S. Bhunia, An automated configurable Trojan insertion framework for dynamic trust benchmarks, in *Design Automation and Test in Europe (DATE)*, pp. 1598–1603 (2018)
14. J. Cruz, F. Farahmandi, A. Ahmed, P. Mishra, Hardware Trojan detection using ATPG and model checking, in *International Conference on VLSI Design* (2018), pp. 91–96
15. J. Cruz, P. Mishra, S. Bhunia, The metric matters: how to measure trust, in *Design Automation Conference (DAC)* (2019)
16. DARPA system security integrated through hardware and firmware (SSITH) (2017). https://www.darpa.mil/program/system-security-integration-through-hardware-and-firmware
17. F. Farahmandi, P. Mishra, Automated test generation for debugging arithmetic circuits, in *Design Automation and Test in Europe (DATE)*, pp. 1351–1356 (2016)
18. F. Farahmandi, P. Mishra, FSM anomaly detection using formal analysis, in *IEEE International Conference on Computer Design (ICCD)* (2017), pp. 313–320
19. F. Farahmandi, P. Mishra, Automated test generation for debugging multiple bugs in arithmetic circuits. IEEE Trans. Comput. **68**(2), 182–197 (2019)
20. F. Farahmandi, P. Mishra, S. Ray, Exploiting transaction level models for observability-aware post-silicon test generation, in *Design Automation and Test in Europe (DATE)* (2016), pp. 1477–1480
21. F. Farahmandi, Y. Huang, P. Mishra, Trojan localization using symbolic algebra, in *Asia and South Pacific Design Automation Conference (ASPDAC)* (2017), pp. 591–597
22. X. Guo, R.G. Dutta, Y. Jin, F. Farahmandi, P. Mishra, Pre-silicon security verification and validation: a formal perspective, in *ACM/IEEE Design Automation Conference (DAC)* (2015), pp. 145:1–145:6
23. X. Guo, R.G. Dutta, P. Mishra, Y. Jin, Scalable SoC trust verification using integrated theorem proving and model checking, in *IEEE International Symposium on Hardware Oriented Security and Trust (HOST)* (2016), pp. 124–129
24. X. Guo, R.G. Dutta, P. Mishra, Y. Jin, Automatic code converter enhanced PCH framework for SoC trust verification. IEEE Trans. Very Large Scale Integr. Syst. **25**(12), 3390–3400 (2017)
25. Y. Huang, P. Mishra, Trace buffer attack on the AES cipher. J. Hardware Syst. Secur. **1**(1), 68–84 (2017)
26. Y. Huang, A. Chattopadhyay, P. Mishra, Trace buffer attack: Security versus observability study in post-silicon debug, in *IEEE International Conference on Very Large Scale Integration (VLSI-SoC)* (2015), pp. 355–360
27. Y. Huang, S. Bhunia, P. Mishra, MERS: statistical test generation for side-channel analysis based Trojan detection, in *ACM Conference on Computer and Communications Security (CCS)* (2016), pp. 130–141
28. Y. Huang, S. Bhunia, P. Mishra, Scalable test generation for Trojan detection using side channel analysis. IEEE Trans. Inf. Forensics Secur. **13**(11), 2746–2760 (2018)
29. H.-M. Koo, P. Mishra, Functional test generation using design and property decomposition techniques. ACM Trans. Embed. Comput. Syst. **8**(4), article 32 (2009)
30. Y. Lyu, P. Mishra, A survey of side channel attacks on caches and countermeasures. J. Hardw. Syst. Secur. **2**(2), 33–50 (2018)
31. Y. Lyu, P. Mishra, Efficient test generation for Trojan detection using side channel analysis, in *Design Automation and Test in Europe (DATE)* (2019)
32. Y. Lyu, X. Qin, M. Chen, P. Mishra, Directed test generation for validation of cache coherence protocols, in *IEEE Transactions on Computer-Aided Design of Integrated Circuits and Systems (TCAD)* (February 2018)
33. Y. Lyu, A. Ahmed, P. Mishra, Automated activation of multiple targets in RTL models using concolic testing, in *Design Automation and Test in Europe (DATE)* (2019)
34. P. Mishra, N. Dutt, Modeling and validation of pipeline specifications. ACM Trans. Embedded Comput. Syst. **3**(1), 114–139 (2004)
35. P. Mishra, N. Dutt, Specification-driven directed test generation for validation of pipelined processors. ACM Trans. Des. Autom. Electr. Syst. **13**(2), 36, article 42 (2008)

36. P. Mishra, H. Tomiyama, A. Halambi, P. Grun, N. Dutt, A. Nicolau, Automatic modeling and validation of pipeline specifications driven by an architecture description language, in *Asia and South Pacific Design Automation Conference (ASPDAC) and VLSI Design* (2002), pp. 458–463

37. P. Mishra, R. Morad, A. Ziv, S. Ray, Post-silicon validation in the SoC era: a tutorial introduction, in IEEE Des. Test **34**(3), 68–92 (2017)

38. A. Nahiyan, F. Farahmandi, P. Mishra, D. Forte, M. Tehranipoor, Security-aware FSM design flow for identifying and mitigating vulnerabilities to fault attacks, in *IEEE Transactions on Computer-Aided Design of Integrated Circuits and Systems (TCAD)* (May 2018)

39. A. Pouraghily, T. Wolf, R. Tessier, Hardware support for embedded operating system security, in *International Conference on Application-specific Systems, Architectures and Processors (ASAP)* (2017), pp. 61–6

40. X. Qin, P. Mishra, Directed test generation for validation of multicore architectures. ACM Trans. Des. Autom. Electron. Syst. **17**(3), article 24, 21 (2012)

41. X. Qin, P. Mishra, Scalable test generation by interleaving concrete and symbolic execution, in *International Conference on VLSI Design* (2014), pp. 104–109

42. K. Rahmani, P. Mishra, Feature-based signal selection for post-silicon debug using machine learning, in *IEEE Transactions on Emerging Topics in Computing (TETC)* (December 2017)

43. K. Rahmani, S. Ray, P. Mishra, Post-silicon trace signal selection using machine learning techniques. IEEE Trans. Very Large Scale Integr. Syst. **25**(2), 570–580 (2017)

44. T. Thomas, A. Pouraghily, K. Hu, R. Tessier, T. Wolf, Multi-task support for security-enabled embedded processors, in *International Conference on Application-specific Systems, Architectures and Processors (ASAP)* (2015), pp. 136–143

Acknowledgments of Copyrighted Materials

Chapter 2 used some materials that are copyrighted by ACM. The definitive versions appeared in the ACM Transactions on Embedded Computing Systems [1] and in the proceedings of ACM Great Lakes Symposium on VLSI [2]. The work is produced under permission granted by ACM as stated by the policy: *Under the ACM copyright transfer agreement, the original copyright holder retains the right to reuse any portion of the work, without fee, in future works of the author's own, including books, lectures and presentations in all media, provided that the ACM citation and notice of the Copyright are included.* Chapter 2 also used some materials that are copyrighted by Springer. The definitive version appeared in the Springer Design Automation for Embedded Systems [3]. The work is produced under automatic permission granted by Springer since the book will be published by Springer. Chapter 2 also used some materials that are copyrighted by IEEE. The definitive version appeared in the proceedings of IEEE International High Level Design Validation and Test Workshop [4]. The work is produced under permission granted by IEEE as we are the authors of the original work and by the policy: *As a member of the IEEE, you are permitted to reuse this content for free except for a $3.50 handling charge in order to obtain a copyright compliance license through RightsLink.*

Chapter 3 used some materials that are copyrighted by Springer. The definitive version appeared in the Springer Design Automation for Embedded Systems [3]. The work is produced under automatic permission granted by Springer.

Chapter 4 used some materials that are copyrighted by ACM. The definitive version appeared in the proceedings of International Symposium on Hardware/Software Codesign and System Synthesis [5]. The work is produced under permission granted by ACM as stated above.

Chapter 5 used some materials that are copyrighted by IEEE. The definitive versions appeared in the proceedings of International Conference on VLSI Design

© Springer Nature Switzerland AG 2020
F. Farahmandi et al., *System-on-Chip Security*,
https://doi.org/10.1007/978-3-030-30596-3

[6] and in IEEE Transactions on Computer-Aided Design of Integrated Circuits and Systems [7]. The work is produced under permission granted by IEEE as stated above.

Chapter 6 used some materials that are copyrighted by EDAA. The definitive version appeared in the proceedings of Design Automation and Test in Europe [8]. The work is produced under permission granted by EDAA as stated by the policy: *Authors and employers retain all proprietary rights in any process, procedure or article of manufacture described in the work, and similarly retain unconstrained rights to publish, print or copy the above work in any form.* Chapter 6 also used some materials that are copyrighted by IEEE. The definitive versions appeared in the IEEE Transactions on Computers [9]. The work is produced under permission granted by IEEE as stated above.

Chapter 7 used some materials that are copyrighted by IEEE. The definitive version appeared in the proceedings of International Conference on VLSI Design [10]. The work is produced under permission granted by IEEE as stated above.

Chapter 8 used some materials that are copyrighted by ACM. The definitive version appeared in the ACM Transactions on Embedded Computing Systems [11]. The work is produced under permission granted by ACM as stated above. Chapter 8 also used some materials that are copyrighted by EDAA. The definitive version appeared in the proceedings of Design Automation and Test in Europe [12]. The work is produced under permission granted by EDAA as stated above.

Chapter 9 used some materials that are copyrighted by EDAA. The definitive version appeared in the proceedings of Design Automation and Test in Europe [13]. The work is produced under permission granted by EDAA as stated above.

Chapter 10 used some materials that are copyrighted by IEEE. The definitive version appeared in the proceedings of International Symposium on Quality Electronic Design [14]. The work is produced under permission granted by IEEE as stated above. Chapter 10 also used some materials that are copyrighted by ACM. The definitive version appeared in the ACM Transactions on Design Automation of Electronic Systems [15]. The work is produced under permission granted by ACM as stated above.

Chapter 11 used some materials that are copyrighted by EDAA. The definitive version appeared in the proceedings of Design Automation and Test in Europe [16]. The work is produced under permission granted by EDAA as stated above.

Chapter 12 used some materials that are copyrighted by ACM. The definitive version appeared in the ACM Transactions on Embedded Computing Systems [1]. The work is produced under permission granted by ACM as stated above. Chapter 12 also used some materials that are copyrighted by IEEE. The definitive version appeared in the proceedings of IEEE International High Level Design Validation and Test Workshop [4]. The work is produced under permission granted by IEEE as stated above.

References

1. M. Chen, P. Mishra, D. Kalita, Automatic RTL test generation from SystemC TLM specifications. ACM Trans. Embed. Comput. Syst. (TECS) **11**(2), article 38 (2012)
2. M. Chen, P. Mishra, D. Kalita, Coverage-driven automatic test generation for UML activity diagrams, in *Proceedings of ACM Great Lakes Symposium on VLSI (GLSVLSI)* (2008), pp. 139–142
3. M. Chen, P. Mishra, D. Kalita, Efficient test case generation for validation of UML activity diagrams. Des. Autom. Embed. Syst. (DAES) **14**(2), 105–130 (2010)
4. M. Chen, P. Mishra, D. Kalita, Towards RTL test generation from SystemC TLM specifications, in *Proceedings of IEEE International High Level Design Validation and Test Workshop (HLDVT)* (2007), pp. 91–96
5. H. Koo, P. Mishra, Specification-based compaction of directed tests for functional validation of pipelined processors, in *Proceedings of International Symposium on Hardware/Software Codesign and System Synthesis (CODES+ISSS)* (2008), pp. 137–142
6. P. Mishra, M. Chen, Efficient techniques for directed test generation using incremental satisfiability, in *Proceedings of International Conference on VLSI Design* (2009), pp. 65–70
7. M. Chen, P. Mishra, Functional test generation using efficient property clustering and learning techniques. IEEE Trans. Comput. Aided Des. Integr. Circuits Syst. (TCAD) **29**(3), 396–404 (2010)
8. M. Chen, X. Qin, P. Mishra, Efficient decision ordering techniques for SAT-based test generation, in *Proceedings of Design Automation and Test in Europe (DATE)* (2010), pp. 490–495
9. M. Chen, P. Mishra, Property learning techniques for efficient generation of directed tests. IEEE Trans. Comput. (TC) **60**(6), 852–864 (2011)
10. X. Qin, M. Chen, P. Mishra, Synchronized generation of directed tests using satisfiability solving, in *Proceedings of International Conference on VLSI Design* (2010), pp. 351–356
11. H. Koo, P. Mishra, Functional test generation using design and property decomposition techniques. ACM Trans. Embed. Comput. Syst. (TECS) **8**(4), article 32 (2009)
12. H. Koo, P. Mishra, Functional test generation using property decompositions for validation of pipelined processors, in *Proceedings of Design Automation and Test in Europe (DATE)* (2006), pp. 1240–1245
13. M. Chen, P. Mishra, Decision ordering based property decomposition for functional test generation, in *Proceedings of Design Automation and Test in Europe (DATE)* (2011), pp. 167–172
14. X. Qin, P. Mishra, Efficient directed test generation for validation of multicore architectures, in *Proceedings of International Symposium on Quality Electronic Design (ISQED)* (2011), pp. 276–283
15. X. Qin, P. Mishra, Directed test generation for validation of multicore architectures. ACM Trans. Des. Autom. Electron. Syst. (TODAES) **17**(3), article 24 (2012)
16. X. Qin, P. Mishra, Automated generation of directed tests for transition coverage in cache coherence protocols, in *Proceedings of Design Automation and Test in Europe (DATE)* (2012), pp. 3–8

Index

© Springer Nature Switzerland AG 2020
F. Farahmandi et al., *System-on-Chip Security*,
https://doi.org/10.1007/978-3-030-30596-3

Printed in the United States
By Bookmasters